COMPARING RURAL DEV

Perspectives on Rural Policy and Planning

Series Editors:
Andrew Gilg
University of Exeter, UK
Professor Keith Hoggart
Kings College, London, UK
Professor Henry Buller
University of Exeter, UK
Professor Owen Furuseth
University of North Carolina, USA
Professor Mark Lapping
University of South Maine, USA

Other titles in series

Sustainable Rural Systems
Sustainable Agriculture and Rural Communities
Edited by Guy M. Robinson
ISBN 978 0 7546 4715 7

Governing Rural Development
Discourses and Practices of Self-help in Australian Rural Policy
Lynda Cheshire
ISBN 978 0 7546 4024 0

The City's Hinterland
Dynamism and Divergence in Europe's Peri-Urban Territories
Keith Hoggart
ISBN 978 0 7546 4344 9

Women in Agriculture in the Middle East
Pnina Motzafi-Haller
ISBBN 978 0 7546 1920 8

Comparing Rural Development
Continuity and Change in the Countryside
of Western Europe

Edited by

ARNAR ÁRNASON
University of Aberdeen, UK

MARK SHUCKSMITH
Newcastle University, UK

JO VERGUNST
University of Aberdeen, UK

LONDON AND NEW YORK

First published 2009 by Ashgate Publishing

2 Park Square, Milton Park, Abingdon, Oxon OX14 4RN
711 Third Avenue, New York, NY 10017, USA

Routledge is an imprint of the Taylor & Francis Group, an informa business

First issued in paperback 2016

British Library Cataloguing in Publication Data
Comparing rural development : continuity and change in the
  countryside of Western Europe. - (Perspectives on rural
  policy and planning)
1. Rural development - Europe, Western - Sociological
aspects - Cross-cultural studies 2. Social capital
(Sociology) - Europe, Western - Cross-cultural studies
I. Árnason, Arnar II. Shucksmith, Mark III. Vergunst, Jo
307.1'412'094

Library of Congress Cataloging-in-Publication Data
Árnason, Arnar.
  Comparing rural development : continuity and change in the countryside of Western
Europe / by Arnar Árnason, Mark Shucksmith and Jo Vergunst.
    p. cm. -- (Perspectives on rural policy and planning)
  Includes bibliographical references and index.
  ISBN 978-0-7546-7518-1
1. Rural development--Europe, Western. I. Shucksmith, Mark. II. Vergunst, Jo Lee. III.
Title.
  HN380.Z9C6185 2008
  307.1'412094--dc22

2008031184

ISBN 978-0-7546-7518-1 (hbk)
ISBN 978-1-138-27253-8 (pbk)

Contents

List of Figures, Map and Tables

Figures

Map

Tables

Notes on Contributors

Arnar Árnason is Lecturer in Social Anthropology in the Department of Anthropology, University of Aberdeen.

Claudio Cecchi is Professor of Rural Developmet at the Faculty of Economics and Director of SPES Development Studies at SAPIENZA University of Rome.

Chris Curtin is Professor in the Department of Political Science and Sociology, National University of Ireland, Galway.

Frances Hannon works as a Senior Researcher with the National Disability Authority, Dublin.

Esko Lehto is a social geographer at the Karelian Institute of the University of Joensuu, Finland.

Andrea Nightingale is Lecturer in Human Geography in the School of Geosciences, University of Edinburgh.

Ronald Macintyre works for the Open University in Scotland.

Torill Meistad works at the Norwegian Olympic and Paralympic Committee and Confederation of Sports, Oslo.

Jukka Oksa is a rural sociologist working at the Karelian Institute of the University of Joensuu, Finland.

Mark Shucksmith is Professor of Planning in the School of Architecture, Planning and Landscape, University of Newcastle.

Susanne Stenbacka is Senior Lecturer at the Department of Social and Economic Geography at Uppsala University.

Karin Tillberg Mattsson is a Research Advisor at the Centre for Welfare Research at Region Gävleborg, Sweden.

Jo Vergunst is an RCUK Academic Fellow in the Department of Anthropology, University of Aberdeen.

Preface and Acknowledgements

Our research was carried out under the auspices of the European Commission 5th Framework Programme (QLK5-CT-2000-00819). We gratefully acknowledge the support of the European Commission. The project was entitled 'Restructuring in Marginal Rural Areas: The Role of Social Capital in Rural Development' and given the name RESTRIM, which occasionally appears here. In this book we emphasise the comparative dimension to the work which we feel lends it special value. We adopted a practice of sharing the primary data between all six research teams and combining it with a series of meetings and study visits in the case study areas themselves. This allowed all the researchers to become familiar with all the case studies and laid the foundations for these comparative studies. Each research team took the lead for one comparative theme and gathered data from all the other teams for it, through a process of data sharing, report writing and regular meetings and conversations. The Irish and Norwegian teams pooled their efforts to explore aspects of the same theme (chapters five and six), leaving the five themes that are presented here.

It was also an important experiment in *qualitative* research collaboration, where we shared stories, experiences and attitudes, rather than what would have been perhaps more easily transferable numerical data. We felt the qualitative material would get us closest to understanding how rural development happens from the point of view of those actually involved in it. And it is this aspect, garnered through careful and sensitive fieldwork, that the academic and policy communities surely have much to learn from. It allows for a questioning of what rural development might mean, and as much attention can paid to how the positive aspects of people's lives can be continued as it is to how change is introduced. The emergence of a common framework for analysis is partly a testament to the efforts and communication skills of the research teams, and also, of course, to the interest shown in the aims our project by the very many informants whom we spoke and worked with across the six case studies.

Montserrat Pallares Barbera, of the Autonomous University of Barcelona, was the Project Observer and contributed many useful comments and reflections, for which we are grateful. Sjur Baardsen was our contact at the European Commission and we thank him for his support. The photographs in Figure 3.2, 4.1, 4.2, 5.1 and 5.2 were taken by Jukka Oksa, and that in Figure 7.1 by Andrea Nightingale, to whom the editors are grateful. The editors and contributors collectively thank all the participants who contributed to the project for the time they spent with us.

Arnar Árnason
Mark Shucksmith
Jo Vergunst

Chapter 1

Introduction

Arnar Árnason, Andrea Nightingale, Mark Shucksmith
and Jo Vergunst

This book describes and compares case studies in rural development in six European countries: Finland, Ireland, Italy, Norway, Scotland and Sweden.[1] Our starting points are many: a old limestone quarry, a new marketing network, and an ongoing debate about public and private service provision, amongst others. We use these to tell the stories of continuity and change in the countryside as the people in case study areas told them to us. We shift between a close-grained description of the thoughts and actions of the people involved and an analytical mode that brings a broader understanding of how rural development has been taking place on the ground. There is one clear lesson: that development happens through social processes, and in particular social networks, that come before and, in some form or other, will last longer than any discrete development project. Understanding development requires a recognition that the dynamic of social continuity and change is key.

But there is limited value in presenting simply another set of discrete cases, and our goal is more ambitious. For it is in the careful comparison of cases that broader and deeper understanding emerges. Each of the stories we were told has its resonances in other areas and other contexts and through following these resonances, certain themes emerge which have relevance wider than the immediate localities. What is presented in this book, therefore, are not the isolated case studies which leave the reader to attempt a comparison between them, but a set of thematic studies which draw on the same core material but explore it in different ways, comparing the cases as they go along. The contributors have been able to do this as a result of close collaboration over three years involving fieldwork and analysis shared between research teams. More will be explained of our methodology within the chapters themselves, but it will be useful at this stage to outline the key themes of our comparison in a straightforward manner.

Networks in development projects (Chapter 1 – 'Networks for Local Development: Aiming for Visibility, Products and Success' by Lehto and Oksa). We start with a comparison of the various networks instigated for rural development studied in the six areas. This theme provides an introduction to the local issues in rural development that the subsequent themes investigate,

---

1   We recognise and are grateful for the contributions of the other RESTRIM researchers to this chapter.

and makes the case that the particular characteristics of the networks are very relevant to the form of development that results. In flourishing regions, networks are able to reorganise themselves around their success stories and enable a sense of continuity through development.

The provision of public goods and public services (Chapter 3 – 'The Process of Building Social Capital in Rural Areas: Public Goods and Public Services' by Cecchi). This theme looks at social capital from an economic perspective, drawing links between social capital, public goods and public services. It explores how different state structures offering varying levels of services and engagement locally can produce different forms and levels of social capital. Where no strong local dynamic exists for rural development, as in our Italian case, the state can take the lead in providing a model for collective action.

Gender, civil society and labour (Chapter 4 – 'Gendered Social Capital: Exploring the Relation between Civil Society and the Labour Market' by Stenbacka and Mattsson). Drawing initially from the Swedish case study, this theme offers evidence to problematise the Putnamian conception of social capital as belonging equally to a 'whole community'. The authors discuss different modes of engagement with voluntary work and the varying production of and access to social capital that can be seen to result, and gender dimensions are particularly significant here.

Cultural identities in rural development (Chapter 5 – 'Identity-building in Regional Initiatives for Rural Development: Comparing Ireland's Lake District and Norway's Mountain Region' by Meistad and Chapter 6 – 'The Role of Identity in Contemporary Rural Development Processes' by Meistad, Hannon and Curtin). Many varieties of social identity exist in the different case study areas and this theme charts their impact upon both specific development projects (Chapter 5) and the overall course of rural development (Chapter 6). How people claim and attribute identities as 'local' or 'expert', for example, influence social relations within development, while the senses of belonging that occur within particular environmental or cultural settings are likely to affect how development processes are understood within a locality. Meistad (Chapter 5) applies this discussion to two projects in Ireland and Norway involving the construction of new regional identities which have very different stories to tell. The central purpose of the theme to problematise the easy association of region, culture and identity that are often made within both development practice and academic commentary.

Networks in environment and landscape (Chapter 7 – 'Using Environmental Resources: Networks in Food and Landscape' by Vergunst, Árnason, Macintyre and Nightingale) This theme focuses on new uses of resources in the environment in the case study areas, the networks involved, and one frequent outcome: new ways of branding rural Europe. Referring to local food development on Skye, the re-use of a quarry in Sweden and a number of other cases, the theme highlights some challenges of branding and also adds to the questioning of the notion of post-productivism as a characterisation of rural Europe. A concern of many people in the case studies is to maintain the productivity of their land, but sometimes in unexpected and non-traditional ways.

The book ends with a set of reflections and conclusions drawn from across these themes. We show that social processes, through networks, are fundamental to development. Our idea of positive rural development, then, entails overcoming the rather staid dichotomies of endogenous versus exogenous development, or infrastructure versus community development. We look to a plurality of cultural identities, a mixing of spatial scales – where a place can be at once local and part of the international economy, and, as we practice here, to the telling and re-telling of narratives of development.

Rural Development in the European Union

The decline in farming as an activity and as a determinant of land use is well known in rural Europe and is the main context for our research on rural, rather than just agricultural, development. Simultaneously, new demands are being made on rural space. In brief, as the post-war consensus came to an end in the 1970s, governments began to look for ways to cut spending, and the Common Agricultural Policy, as the largest expenditure of the EU, was seen to be in line for cutbacks. Agricultural production had outstripped demand, while at the same time many rural areas were suffering continued population loss, lack of services, economic underperformance and environmental degradation. Southern enlargement of the EU put pressure for CAP reform and pushed new ideas of territorial development and 'rural', as opposed to 'agricultural' policy, onto the agenda. The recent eastern expansion of the EU has heightened concerns over spending on farm subsidies. A succession of EU policy documents, starting with 'The Future of Rural Society' (European Commission 1988) have been designed to implement a territorial approach to rural development, which can also be seen as an attempt to move from a 'sectoral' towards an 'integrated' rural development strategy (Shucksmith et al. 2005).

Alongside such political and economic changes, there are a set of social and cultural factors that have been influencing how environments are used in rural Europe. Marsden (1998, 15) noted, for example, that 'new demands, for "quality" food production, public amenity space, positional residential property, areas of environmental protection, and for the experience of different types of rural idyll or urban antithesis are now much more entrenched in rural space'. These demands have arisen not just as a result of population movement from urban to rural areas, although this counter-urbanisation has for some time been important dynamic in many rural areas (Fielding 1982), but from a broader, European-wide rethinking of the uses of rural areas. Food scares, such as BSE, have drawn attention to the quality rather than just the quantity of food produced in rural Europe. Mass-membership environmental movements in Europe also signify a concern over damage to the natural environment, and this has been translated into a number of EU policies, including the Birds Directive of 1979 and the Habitats Directive 1994. Finally, 'endogenous' rural development strategies have become widespread, whereby people in rural areas themselves attempt to identify and utilise local resources for local benefits (Shucksmith 2000).

The changes in rural Europe from an agricultural to a broader social and economic base have for some time been theorised as the 'post-productivist transition' (Ilbery 1998). According to this, consumption is becoming more important in rural areas. The rise of tourism, different forms of land-use competition, and concerns over the environment are held to be linked together as part of a new type of post-productivist rurality. Debate continues as to how applicable this model is across Europe (Hoggart and Paniagua 2001; Marsden 2003; Ward et al. 2008). But if there is something in the argument, we need to consider both what is happening to European rural space and how it is happening.

Agency in Rural Development: Communities, Individuals and Networks

We now need to be clear about how we conceptualise development itself. We are using the term in a broad sense to refer to processes of social change and continuity, encompassing both planned social change, such as 'development projects', and the everyday ways of life that people attempt to continue and improve by their own efforts. The research presented in this book tracks the means by which these often rather different processes play out in particular case study contexts. We often return to the themes of continuity and change, looking at, for example, the ways that development projects are often relatively short-term affairs and sometimes contrast with other efforts simply to continue valued ways of life. The constant search for newness and innovation in rural development projects – where finance is available for business expansion but not maintenance, for example – marks out a distinctive discourse of development related to the ideals of progress that have been subject to the broader critique of modernist development (Escobar 1995; Gardner and Lewis 1996; Abram and Waldren 1998).

Kovách and Kucerova have suggested that rural development in Eastern Europe is increasingly subject to 'projectification' (Kovách and Kucerova 2006). They suggest that rural development elites may be increasingly pursuing development agendas that reflect their own intellectual and economic interests, partly as a result of decentralisation of budgets to rural and regional levels. The contributors to this book similarly find it useful to maintain a critical approach to development. From our perspective we could disagree with the idea that decentralisation is the problem, since in many of our case study areas, such as Scotland, we have seen a tendency towards the loss of autonomy over budgets and thus economic development decisions within localities, suggesting that the entrepreneurial ethos is in fact perpetuated at larger scales. The specific trends of Eastern European rural development are not within the scope of this book, and our research took place before the A8 accession. But Kovách and Kucerova usefully distinguish development outcomes between on one hand the revitalising of civil society and on the other the instrumental use of funds for limited interests. This certainly chimes with our research in rural Western Europe, as we find some success stories and some – while not necessarily finished – that are not characterised thus by participants in them. It also indicates that EU regional funding may in some ways mirror the 'Third Way' approaches of other liberalised democracies (Kendall 2000; MacKenzie 2004), where policy spaces are created in which 'policy entrepreneurs' can operate. In this book we take on board these critiques of development and keep a broad view in our case study areas. We want to open the relationship between formal development projects and ongoing social life to critical evaluation.

The next conceptual issue, once we accept the importance of understanding the agency of development professionals, is in how best to attribute agency to other kinds of people – such as individual residents in our case study areas – or to 'communities' within studies of rural development. We meet the problem of whether it is better to conceive of the 'individual' or the 'community' as the key actor in development processes, and this is relevant both methodologically and analytically. This leads us into some of the key themes for the book.

The question of individual and community roles has been important in studies of both formal development projects and of local social life more generally. Shucksmith raised the issue in 2000, as part of an evaluation of LEADER in the west of Scotland. In some attempts at endogenous development (i.e. development that emerges and uses resources from within a place, broadly speaking) easy assumptions are often made about 'the community', and unequal power relations within the group and differential participation in development processes are often ignored (Shucksmith 2000, 209–10). 'Community development' runs the risk of reifying traditional notions of rural society as homogenous and bounded, despite prolonged social science attention towards rural class structure and other internal differentials (e.g. Newby 1979; Milbourne 1997). 'Community' is often equated with 'territory', with the result that these questions of internal social differentiation may not be asked. If regional territorial identity is becoming more significant in European rural development (Ray 1999), there is the possibility that symbols of collective identity paper over more specific issues and hide inequalities in power, wealth and access. As Shucksmith put it, from the point of view of people within these communities, 'their (individual) capacity to act will be diminished by such approaches to (collective) capacity building, and it is unclear prima facie whether this will increase or decrease inequalities in society' (Shucksmith 2000, 10). This is a critique of the concept of community from the point of view of the individual, whereby attempts to build the capacity of the collective in determining the course of development can end up excluding or disempowering certain individuals from it.

The idea of community therefore appears as a symbol that is available for manipulation and use through discourse and action. Cohen (1985; 1987) explores the 'symbolic boundaries' of community in detail, showing how common distinctions between 'us' and 'others' are maintained through small acts of speech, narrative and everyday action. Having long since lost its status as an inviolable empirical fact (Bell and Newby 1971), community-as-symbol alerts us to the production of the commonalities of a social group and territory on the one hand, and the production of difference to others on the other. In academic terms, if not always in lay discourse, 'community' has long since lost any veneer of neutrality. In recognising 'community' in this way we destabilise it and call into question those regional development initiatives that place it at the centre of rural policy. It is important to scrutinise the political processes through which 'communities' are constituted and social capital attributed to them, as well as the contexts within which notions of community and of individuals belonging to a collective are formed.

It is notable, however, that the idea of the individual is rarely subjected to similar critical scrutiny. If we need to be careful about how we use the term community, because of its symbolic construction, what should we make of the 'individuals' to whom it is often opposed? On principle it seems unrealistic to identify one half of a dichotomy as being constructed and thus problematic, while accepting the other as natural. To be fair, Bourdieu's concepts of social capital and symbolic violence, used by Shucksmith (2000) to critique the concept of community, must be understood as part of his broader theory of practice in which individuals are viewed as internalising the social field through their habitus. Bourdieu is highly critical of rational action theory. But in development studies we often unpack one side of the dichotomy without paying so much attention to the other. In practice this means we have focused on the problem of community without making similar efforts to uncover the symbols or other insinuations that result from analytically constructing people as 'individuals'.

Individuals are the unit and subject of much social science research. Often, we as researchers then build up the 'social' dimension by collating responses of individuals: reading off the collective attitude or activity from the aggregated units. The dichotomy of individual and society (or community) then emerges through this process of building up from the former to the latter. On the other hand, critiques of the notion of community often carry out the same process in reverse, pointing out the inadequacies of referring to the collective when individuals are overlooked as a result. The question of how to attribute social capital to communities or individuals faces this problem. While there are many subtle analytical critiques of community (such as Cohen's), there is a need to apply the same rigour to concepts of the individual. Discussions of social capital could be enlivened by understanding how the actions of people, including those amongst whom we carry out research, are inherently social and performed, being continually placed within networks of social relations, as has begun to happen (Murdoch 2000). Our goal then is not to switch between the analysis of a separated, rationally-minded individual and a super-organic or aggregated collective but to investigate the ways in which the attitudes and actions of people are meaningful in themselves, acting out the sociality (the social process) which is often sought at the level of the community. Reconsidering of the terms of the question – what we actually mean by 'individuals' and 'communities' – should allow more nuanced reflections on the quality of social life in the case study areas. We study networks in this book to track how people come together and move on from particular social groupings.

Researching Rural Development in a Multidisciplinary and International Framework

Our research encompasses ideas about rural development, networks and social capital that often hinge on the issues of individual and community analysis outlined here. Like the other concepts, how social capital is conceived of depends on one's grounding in the social sciences. One way of distinguishing between concepts of social capital is to contrast an approach inspired by Emile Durkheim, focusing on the social solidarity that produces the 'glue' to bind members of a society (e.g. Putnam 1993), with a Weberian approach that examines the divisions within society, and particularly the mechanisms by which individuals are able to accumulate various sorts of capital (e.g. Bourdieu 1984). In the latter, social capital relates to the possibilities that individuals have for establishing or cementing themselves within social networks, in order to attain personal 'advancement' or maintain social distinction. The former, by contrast, emphasises the collective means of improvement that can be undertaken by groups that are able to operate successfully together. As a large and multidisciplinary team of social scientists, it was inevitable that people within our group would conceptualise social capital in different ways. While these did not necessarily correlate exactly to the two simplified models, the models were useful in understanding where our various interests came from. What also became clear was that different models of social capital would be most relevant in analytical terms in different empirical circumstances. One issue we returned to periodically was the extent to which we should maintain coherence in our overall approach to social capital.

Our research teams followed a diversity of approaches to social capital. We refrained from imposing a single, all-encompassing definition of social capital, partly so that the teams could follow their own theoretical interests, but also, and perhaps more significantly, so that the differences between the research areas and themes could be explored in a more sensitive way. For example, Cecchi writes in Chapter 3 that 'social capital is the result of the use of resources – that might have been used in a different way – whose benefits can influence the performance of the community for a long period of time'. This frames his subsequent interest in the relationship between the provision of public services and local social structures. Social capital is understood as a phenomenon that can be invested in, as public services can provide a model for collective action that can be adopted by local actors. The collective aspects to social capital are emphasised by this approach. Similarly, Lehto and Oksa ground Chapter 2 in an idea of social capital as the ability to collectively exploit a set of resources, which in themselves can include group working. In the same way that Cecchi discusses how social capital is formed through the provision of services and yet is also required in the negotiation of those services, Lehto and Oksa argue that local resources are transformed through the presence of positive social relations: these networks are both a resource and a way of using other resources.

These collectivist approaches reflect a concern with the cohesiveness and 'functioning' of society in a Durkheimian sense. Lehto and Oksa, however, also drew their conclusions from a detailed discussion of networking in our case study areas. For us, the extension of the concept of social capital into studies of networks was a significant step. Networks can be seen to emerge in relation to particular issues. The events described by Lehto and Oksa were often of a temporary nature, bringing together people in various formations, who often had differing interests in the event. Other parts of our research also used networks as a key theoretical idea, describing social networks involving identity (Meistad in Chapter 5, Hannon and Curtis in Chapter 6) and the environment (Vergunst, Árnason, Macintyre and Nightingale in Chapter 7). These 'network' accounts resulted in renditions of social capital as being highly tensioned, where local development trajectories may form and split, or diverge and converge with other local, national or international paths. The ability to turn a network towards a particular point of view or goal is of interest here. In doing so, the researchers use a model of 'society' that is some distance from a clearly bounded and homogenous entity. Instead, their focus on social relations bears more in common with studies of agency that come from the work of Max Weber (Weber 1978), and Bourdieu's notion of social capital as incorporating the ability to exert influence over others is also relevant here. Importantly, most of these studies emphasize the performance of identity and power within specific contexts. It is these performances that link the individual and the collective.

In Chapter 4, Mattsson and Stenbacka maintain that qualities of relationships between groups and individuals are important in development processes. They perhaps come closest to reconciling the two approaches to social capital outlined here, arguing that while active voluntary associations can contribute as a whole to development processes, connections between actors on an individual level need to be understood in order to ascertain specific effects on different social groups. Age, gender and place-identity (in other words, perceptions of belonging to a place) are the main categories that they find can be factors in determining the relationship between voluntary activities and engagement in the labour market. Their typology of outcomes from this relationship at the end of their chapter shows an unwillingness to prescribe a single way of understanding social capital. Here again, social capital emerges more as a practice or process than a pre-existing quality.

Applied to the level of our research in its entirety, this perspective goes some way to overcoming some difficulties with the concept of social capital (Fine 2001; 2003). The under-theorised nature of social capital means that it is hard to know exactly what is being referred to. By gathering qualitative data on rural development processes and broader associational practices within the case study areas, we have produced accounts that are grounded within wider social science theory than simply that of social capital. That the accounts differ in their particular theoretical slant reflects in equal measure the empirical diversity of the case study material and the theoretical interests of the researchers, both personally and as a group. The choice of themes came about through the relating of the empirical to the theoretical concepts – a flexible process that did not happen once and for all, but was ongoing through the life of the research. This is why we characterise our approaches to social capital as stories: narrative explorations of process and performance, rather than cause-and-effect explanations of difference. While we do not attempt to refute Fine's extensive critiques of social capital here, we hope that emphasising the contexts within which the particular networks that we are interested in have arisen will allow us to maintain a complexity and depth in our explanations.

The notion of social capital, as we argue in more detail the conclusion, is best understood as a metaphor for the qualities of some social relationships that allow other benefits to be secured through them. Such qualities may include trustworthiness or confidence, or values based in kinship, co-residence, shared work, or other shared experience. So social capital should not be seen as belonging to either communities or individuals but is an aspect of social relationships through which both communities and individuals can be constituted. This approach (which we would hesitate to call a definition) draws attention on the one hand to the social relations and networks in which productive relations are based, and, on the other, to the exploration of metaphor, which suggests narrative modes of research and writing that emphasise process rather than simple cause-and-effect. We seek to unpack the variety of possibilities within the concept of social capital in much the same way as we explore the variety of rural development practices in Europe. Furthermore, this framework opens the way towards a discussion of individual and community in the terms outlined here. Diversity in rural development practice requires that we try and understand the particular ways that the objects of development – communities or individuals, for example – are themselves produced through discourse and action. This is intended not as a counter balance to previous attention towards 'community' in rural development analysis, but as a way of re-thinking the whole idea of a community-individual dichotomy.

Methodology and Case Study Areas

The methodology for this project was worked out collectively and adapted as the research continued. While the importance of producing adequate comparisons between the case studies was paramount, the qualitative research techniques were flexible enough to allow the unique themes in each field area to be followed up. Once the broad parameters of the project had been set, each of the six research teams began the case study investigations by producing a context report on the overall economy and population of their region. Fieldwork in the case studies then took place over the period of approximately one year, and involved a directed questionnaire study, informal and semi-structured interviewing with both key informants and ordinary inhabitants, and attendance at relevant public meetings, associational activities and other occasions as they presented themselves. Each team then produced a report on their own fieldwork in which ideas for the comparative themes began to emerge, and these were discussed and refined during regular meetings of all the research teams. Each comparative theme was written up into the chapters presented here through sharing the prior reports, through conversation and comment between the teams, and by drawing on the field visits we made to most of the case study areas as part of our project meetings.

The case study areas are described in detail when they become relevant to each of the themes. Nonetheless it will be useful here to provide some background information to each. Our interests are primarily in the remote rural parts of Europe, but we were concerned not so much with the geographical spread as with the variety of social networks that are relevant to development in each of them.

Finland

Finland is one of Western Europe's most rural countries. Its settlement structure is characterised by the low population density, only 16 inhabitants per km2, the lowest in the European Union. The 'rural question' has always been at the top of the state agenda. The central problems of Finland's rural areas include loss of population and decrease in the number of services and jobs available. The concentration of people and operations in growth centres is related to the ongoing transformation of the economic structure, a process that started later in Finland than in many other countries.

Our Finnish research area, the Kainuu region and the Sotkamo municipality, is situated on the north eastern edge of European Union at the border of Finland and Russia. The Kainuu Region has about 91,000 inhabitants and its average population density is only 3.8 persons/ km^2. The capital of Kainuu Region is the city of Kajaani. Natural resources are important for the regional economy of Kainuu. The role of forest sector and agriculture has traditionally been significant, although it is nowadays going through a thorough restructuring

Map 1.1   Location of case study areas

process. The landscape of Kainuu offers possibilities for recreation and tourism. There is plenty of water and 9,500 kilometres of lake shore. The region has implemented several nature protection measures and there is a regional plan for sustainable use of natural resources.

In terms of net migration and the rate of unemployment Sotkamo municipality is one of the least suffering municipalities in the Kainuu Region. The public image of Sotkamo is favourable and Sotkamo is seen as successful in comparison with neighbouring areas. The image of its Vuokatti landscape is well known, and Sotkamo's service sector has been innovative. In Sotkamo one finds very positive self-awareness, strong motivation for local cooperation and one of the top baseball teams in Finland.

Ireland

Even more so than Finland, Ireland has undergone rapid transformation. The 1990s were characterised as the period of the Celtic Tiger, which brought dramatic growth to the Irish economy. Whilst growth has taken place and was rapid, it did not spread itself evenly and many disparities emerged, both between rich and poor and the more and lesser-developed regions of the country. With the exception of Galway city, growth has been much slower in the Western Region, particularly in the counties of Mayo, where the Irish study area is located, Roscommon, Leitrim and Donegal.

The location of the Irish study area is interesting as it is in some ways very isolated through its poor infrastructure, large lakes and mountainous areas, yet it serves as a commuting area to employment in Galway. Like the rest of the Western Region, it is in receipt of Objective 1 funds and there is a wide range of funds available to redress the regional imbalances in development. Such funds are not channelled directly through the government as in other EU countries, but rather specialised and relatively independent development agencies have been established to administer the funds and promote community development in the localities in which they work. Who accesses such funds and how they do so is an important question and that has been raised in several studies and evaluations of the LEADER programme. Civil society plays an important role and it has been suggested that group membership and established social contacts are often prerequisites for gaining access to such funds. The differing roles men and women play is also of interest as there is an overwhelming majority of men in higher-ranking decision making positions, yet more women join and are active in groups. There is also a sense in the locality that it is often the 'same faces' that one sees and there are those who feel excluded. Multidimensionality is an important feature of both the economy and the local culture and people are expected to play a variety of roles. This leads us to the question of how local elites function and whether networks in one sector (e.g. sport) lead to better networks and outputs in other sectors (e.g. business).

Italy

The Italian area of study, the province of Grosseto, covers the northern part of the so-called 'Maremma Tosco-Laziale', a large area stretching from the north of Rome to Livorno in Tuscany. Grosseto is the most southern of the nine provinces of Tuscany. It takes its name from the main town in the area, and includes 28 municipalities. It was chosen as a study area because it was seen as lying at the physical and cultural border between modern European metropolitan areas and old traditional rural communities. It has a history of underdevelopment yet solid historical ties to the developed areas due to its geographical location in between the centre and north of Italy. Also, and most importantly, it is undergoing powerful economic and social transformations.

There is clear evidence that the agricultural production of the Grosseto province has represented a very important part of the food supply for Florence, Siena and Rome ever since the Renaissance period. The province of Grosseto is considered the largest agricultural area in Tuscany. This district has depended largely on the agricultural sector and only in recent times has some attention been directed towards tourism, which has become an important income-generating activity in the western part of the province. The transformation in the local economy of Maremma has taken the form of a dramatic reduction in the size of the agricultural sector, both in terms of employment and of the share of local value added. The urban system is centred on the city of Grosseto, which has acquired the characteristics of an urban area over the last decades, when it has grown both in size and in economic importance.

Norway

The Norwegian study area, the Mountain Region (in the area of Røros), has an internal economic, social and cultural heterogeneity that is representative of Norwegian rural areas. It contains a variety of communities facing different development problems connected to the growth of the rural 'late modern economies', as well as a very wide spectrum of strategies chosen in order to cope with these successfully. What distinguishes the Mountain Region from other similar regions, however, is its relative success in restructuring its economy during the last decades and, as a consequence, in maintaining its population level. There are more jobs in the region today than 20 years ago, and there are almost as many inhabitants living there as in 1986 – a situation very rare for remote rural areas in Norway.

During the last decades the region has undergone profound changes. This is observed most clearly in the local labour market, where the two traditional cornerstone activities of the region – mining and agriculture – have been replaced by the expanding service sectors. Socially and culturally the Mountain Region communities have changed profoundly. In brief, these changes may be summarised as levelling out traditional differences between rural and urban regions, bringing a 'national' culture to the 'periphery' (see Eriksen 1997). At the same time the region's relationship to and dependency on various 'urban', national and global processes have become more transparent.

Scotland

The Scottish study area of Skye and Lochalsh is located on the western side of the Highlands and Islands region. The Highlands and Islands covers 39,050 km², yet the region has less than 10 per cent of the Scottish population with around 440,000 residents, giving a population density of an average of just 11 people per km². The low population density is attributable to the history of feudal land ownership – a few large land owners controlling the vast majority of the land

(Wightman 1996) – and a scarcity of arable land. The land ownership situation is beginning to change with recent community land ownership initiatives (Sellars 2006). As the name suggests, much of the area is of relatively high altitude (up to 1,250m) including the famous Cuillin Hills on the Isle of Skye. The region had a sudden and dramatic decline in population during the nineteenth century when tenants were dispossessed of their land in many parts of the Highlands and Islands. Very few communities have recovered their populations since that time and the region as a whole has approximately 48,000 fewer residents in 2006 than in 1851.

Historically the Highlands and Islands economy has been dominated by agriculture, forestry and fishing in addition to public sector jobs. The small scale tenant farmers, or crofters, have never held enough land to support their households and particularly in the face of declining prices for sheep, rural households continue to rely on a diversity of economic activities. Tourism has been an important part of the economy for decades and continues to be one of the growth sectors. Over recent years renewable energy has become a focus in the Highland and Islands, with a large scale (250 turbines) wind project moving forward in the Outer Hebrides, and smaller proposals for Skye. It is within this context that Skye and Lochalsh has been comparatively successful at attracting new jobs and residents. Skye in particular has been seen to have a 'magic' that attracts many tourists and in-migrants.

While the economy of Skye is growing, that growth is not necessarily occurring in sectors that provide long-term, non-seasonal and relatively high paying employment. The public sector is an extremely important source of professional jobs, particularly for women, but with the reform of local government in 1996, some of these jobs have been lost. The labour market shows strong differentiation along gender lines with more men involved in the traditional sectors based on natural resources (agriculture, fishing, forestry, etc), finance and manufacturing. Women dominate jobs in the tourism, public administration, education and health sectors, but men hold most of the leadership and management jobs with these sectors. Both tourism and natural resource sectors are largely seasonal, and unemployment increases during the winter months for both men and women.

Sweden

The Swedish study area of Leksand and Rättvik, in Dalarna, represents a common type of small scale farming areas, where farming traditionally was combined with subsidiary industries. Still, while most rural municipalities in Sweden have experienced a continuous population decrease, Leksand has not only kept its population but has during some periods even increased it. The population development over the last 50 years has not, however, been even. There have been periods when the municipality lost more people than was counterweighted by in-migration. The neighbouring municipality Rättvik, which is characterised by similar economic and historical conditions, has not experienced an equally favourable trend but has been able to attract in-movers to some degree.

Today, when the importance of the agrarian sector has diminished, Leksand and Rättvik are still characterised by small scale enterprises and people earning a living by a combination of different activities. The timber industry, that has a long tradition in the area, is still thriving. So is the tourism industry, and this area was among the first in Sweden to be discovered and exploited by tourists. Although Leksand and Rättvik are characterised by quite diversified economy compared to many other rural municipalities, there are some restrictions. For women the economy is far less diversified. The majority of women in the area, like in the rest of Sweden, work in the public sector. As the public sector has been subject to down-sizing during the last decade, women's dependence on this segment of the labour market has made them vulnerable.

The labour market and the economy seem, however, not to be the most important determining factor for people who are choosing to move to the area. What is of greater attraction is the quality of life people hope to find in the area. People move to the area because it is a good place to raise children, because good housing can be found at reasonable prices, and because it is an area that still lives up to the image of the traditional Swedish countryside. A central component for the identity of both the municipalities and individual inhabitants are the closely knit village communities, *byalagen*, that are meeting places for networking and therefore important sources for generating social capital.

Chapter 2
Networks for Local Development: Aiming for Visibility, Products and Success

Esko Lehto and Jukka Oksa[1]

On a very general level we understand rural development as a process which brings about a more diversified economy and improved quality of life. Local development actions can be understood as processes where local actors, including persons, groups and institutions, try to mobilise various kinds of local and external resources. In this chapter we are focusing attention onto those networks from the six case study areas (introductions for which can be found in Chapter 1) that have participated in significant events or turning points of local development. Table 2.1 lists the networks used as the basis for the observations in this chapter.

Framework for Comparison

Our general comparative framework consists of the following concepts. Local practices are the starting point for development. Local practices together with the local resources, such as earlier forms of physical and human capital, are the main sources of continuities with the past, defining the given strengths and weaknesses of locality. The prevailing forms of regulation and administrative institutions set the context for the actions of the local networks. This context includes different positions in local fields of power and decision-making. Local resources are accessible for local actors and external resources may

1   The authors would like to thank all the people in Sotkamo who gave their time to this research. The authors who work in Joensuu want to thank the Kajaani Development Centre of the University of Oulu for various kinds of support and help. Its director Pentti Malinen had the responsibility of the official Finnish partner of the RESTRIM project representing the University of Oulu and making the necessary administrative decisions. Ms. Raija Koskelo helped in conducting the questionnaire survey and interviews in Sotkamo. The authors also acknowledge the contribution of the Finnish RESTRIM National Advisory Group, whose backing and many-sided expertise has been very valuable.

be accessible through network connections, depending on the nature of the network interactions.

Networks, in this framework, are established sets of actors with regular connections, through which actors gain access to some resources for other members of the network. Opening up access to resources may be based on mutual agreements, perceived common aims, trust or common interests. The potential extent of these accessible resources can be seen to be one aspect of social capital. Social capital is constructed and used in this practical functioning of the network (Falk and Kilpatrick 2000).

The case studies carried out in the six countries threw up a good number of networks and associated activities. The networks for development are by no means identical to each other. They vary in composition (of social actors) and scope, in their aims, their time-span, and their practical outcomes. In our comparison we are focusing on three important aspects of the functioning of a network: 1) *the mechanisms of local development* produced or used in the networks; 2) *the scope of resources* that the network opens for its participants; 3) *the evolution of the quality of network connections*. Does the network accumulate experiences and create improved ways of cooperation or otherwise new concepts of local practices? A single project with a short-term time limit may not itself be able to accumulate many lessons, but it may be a part of a longer process of local learning. At its best, a network for development may evolve into the strategic force that is able to shape common ideas about the future of the place and to convince the people about the actions and commitments needed to get there. These three aspects of networking are simultaneously three different but interwoven parts of the process of networking. They all influence each other and together they determine the significance of the network for local development.

We illustrate our framework in Figure 2.1. The circle in the middle represents networks in action, accessing and combining both local and external resources and working in the context of administrative and political structures of the given place. The networks have also connections to the prevailing local development practices and conventions. The mechanisms of local development connect the results of the network to the more general development of the place.

The successes of the reported networks have been of very different kinds and degrees. There are, naturally, differences in kinds of outcome between single projects with a limited time span and of longer-term activities consisting of sets of projects. We try to discuss both of these – single projects and longer changes – in the same framework, by focusing our attention on the mechanisms of change that connect network activities and local development together. Even one short project may be able to instigate a new mechanism of local change (such as a new partnership or a new form of cooperation), and larger changes may also be discussed in terms of the mechanism or the set of mechanisms that has brought them about.

Table 2.1 Networks for development in the case studies

Country	Network case	Location	Activities
Finland	Sotkamo Dairy	Kainuu Region, Sotkamo municipality	An attempt to create new milk product brand and distribution network; produced visibility but failed because of hard competition
Finland	Vuokatti Tourism and Sports	Kainuu Region, Sotkamo municipality	Sport and tourism marketing, joint high visibility Vuokatti brand
Ireland	Making of Lake District	Counties of Mayo and Galway (Ballinrobe town in the middle of the region)	Regional visibility by launching a development enterprise, which organised events and projects
Italy	Maremma food and tourism	Southern Tuscany, Amiata Grossetano and Colline Interne (regions of 10 municipalities)	Reinventing Maremma as a regional brand for tourism and product marketing: more market visibility
Norway	Building Mountain Region	Edges of the Counties of South-Trondelag and Hedmark	New regional unit of cooperation of municipalities in periphery; visibility as a good place to live and work, lobbying for public funds, local newspaper, tourism marketing
Sweden	Sponsoring Leksand Ice-Hockey	Dalarna Region, Leksand municipality	The visibility of top league ice hockey team is used for attracting enterprise contacts (ice-hockey and business marketing)
Sweden	Summer-time Opera at Dalhalla	Dalarna Region, Rättvik municipality	Utilising of old quarry as a summer-time concert arena with high visibility
Scotland	Bed and Breakfast Network	Scottish Highlands, Skye and Lochalsh	Region's visibility is used for joint tourism marketing
Scotland	Horticultural development association	Scottish Highlands, Skye and Lochalsh	Endogenous development by recognising the transfer of local skills of horticulture
Scotland	Local Food Link	Scottish Highlands, Skye and Lochalsh	New logistic innovation for local food distribution; part of national Food Futures project
Scotland	Skye and Lochalsh Marketing Group	Scottish Highlands, Skye and Lochalsh	Launching of a new local brand 'Skye and Lochalsh Pebble'
Scotland	Wind Farm Controversy	Scottish Highlands, Isle of Skye	Struggle against and for the plan to build wind turbines

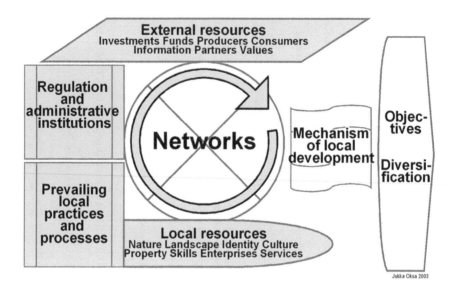

Figure 2.1   Framework for comparing networks for local development

We have divided the 12 networks for local development that we found in the case study areas into three main groups. Firstly, those networks that have started to increase the *visibility* of a relatively unknown place (by which we mean raising its profile), secondly those aiming for *new products* (including services and cultural products), and thirdly those *reorganising* existing activities around an existing successful product or phenomena. These three mechanisms seem to be like three steps following each other. A region starts with measures to increase its visibility, which gives a boost to both new image and new identity. After that the region may start turning the regional visibility into product visibility, of which there are many examples in food products and tourist services. Good visibility may also be a resource in attracting seasonal or permanent residents or enterprises. When a place has developed strong visibility with a help of a success story of any kind, there arises an initiative to reorganise the locality around this new success factor.

Networking for Visibility

Improved visibility of a place is important in many ways, in terms of influencing decision-making, attracting new inhabitants and enterprises, and in particular, visibility means growth in tourism. This seems to be the most common factor in the success stories we found. During the recent decades tourism has been the fastest growing, and sometimes the only growing industry in rural areas (Hall and Page 1999). But for rural people changing from exploiting natural resources (agriculture, forestry, fishery, and hydropower) into serving people (coming from other places to find a unique experience) is a huge change. The core skills needed are different. Struggles with the forces of nature have to be replaced with the skills of interacting with a visitor. In all the case study areas the local people are unanimous about the value of the natural surroundings of their place. However, the social configurations of how to build sustainable jobs and income differ. All the networks want to increase the visibility of their locality, and in some ways all of them have succeeded. However, some networks seem to have gone further down the road than others. The comparative issue could be formulated in the following way: how have networks of local development improved the visibility of their locality, and what are networks doing with the improved visibility?

Trying to Become Visible

We start with the three networks that have taken steps to create visibility for a new area. These networks are: 1) the Lake District network that tried to identify and promote an area in Ireland; 2) the Maremma network in Italy, that is creating a new brand name for tourism and food products of the area; and 3) the Mountain Region network of remote municipalities located on the peripheries of two counties of Norway. These three networks have in common the starting point that they are creating visibility for a newly constructed territorial entity. Therefore, we can simultaneously follow both the process of building identity for oneself and an image for outsiders.

The Irish Lake District came into being when a community development company, the Lake District Enterprise (LDE) was formed in 1997 (Hannon et al. 2003). The area is somewhat isolated because of its poor infrastructure, large lakes and mountains, yet it is only 30 minutes drive from the rapidly growing city of Galway in the west of Ireland. Because of its little publicised natural beauty it has been called 'one of Ireland's best kept secrets'.

Lake District network (Ireland)

Initiated by business actors in Ballinrobe town, non-profit community development enterprise Lake District Enterprise (LDE) was founded to promote the economic, cultural and social development of the region. The LDE started to establish networking structures at the Lake District level and to promote the area as an ideal place to live and work, as well as to advertise it as a leisure and tourist destination.

Achievements: Support to enterprises and many project activities in area development, tourist business plan, rural tourism, computer training, basic infrastructure, community development.

Lake District marketing was not based on an already existing regional identity. However, there were expectations that common identity could be based on the family, school, occupational and cultural networks that radiate throughout the area.

In the Irish Lake District local business people did get a lot of positive visibility through various local events and happenings and supported the network through their membership fees. There was also some external financing for the projects but the support was not strong enough and long enough to establish a permanent institutional structure that would continue the work on a more permanent bases. In principle the development enterprise could have been such a structure, but it depended too much on the expertise of one project manager, who even changed during the process. In spite of a series of good events the networking process could not reach more continuous mechanisms of development than the single projects.

The process of community creation has increased individual and collective self-awareness by creating new opportunities for dialogue and by identifying common goals. At the start of the LDE initiative communities were coming together in different events, such as festivals and fairs, but later financial difficulties left LDE without a manager. In addition, it appears that there would have been more external funds available for additional projects if more local resources had been involved.

The Italian case study area of Grossetto has been participating in a marketing campaign of Maremma foods and tourism services (Cecchi 2003; Cecchi and Micocci 2003). The campaign aims to revitalise the idea of 'Maremma' with the mobilisation of local enterprises and some public resources from the administration of the Province of Grossetto. The project called 'Maremma: the rural district of Europe' is organised as a part of the rural regional plan for the Southern Tuscany Region. The Grossetto area consists of seven small municipalities located at the slopes of Amiata Mountains and of three agricultural municipalities located around Sorano. The resources of the municipalities are very limited. The area is remote from industrial centres. Although its soil has unused reserves of minerals, industrial development has been limited. Tourism to the region has developed slowly, too, and the region has constantly been losing population through out-migration. There are some volunteer organisations and networks that try to tackle social and economic issues around development, such as youth cooperatives, sport associations, cooperatives for the assistance of the elderly, and the local Chamber of Commerce. The field of civil society however seems to be rather fragmented both in terms of sectors and in terms of localities.

Maremma marketing network (Italy)

Actors are individual enterprises in tourism and in food manufacturing, province of Grosseto, local municipalities, and local cooperatives.

The aim was to start the Maremma territorial development process in rural tourism and local food specialities.

Achievements: Coordination of the marketing of tourism and local food specialities with the help of the rural programme of Maremma and local development projects.

In Ireland and in Italy one may note the absence of strong local government that could take the role of coordinating and pushing forwards some strategic projects. This is evidently due in part to the structure of local and regional administration in these countries.

Mountain Region Council (Norway)

Cooperation of the leadership of seven municipalities located in the periphery of two counties of South-Trøndelag and Hedmark; also a district newspaper announced itself to be a herald of the Mountain Region. The council leadership consisted of head administrator, the mayor and one political representative of the opposition party from each municipality. In the beginning there were eight municipalities, then nine, and finally seven municipalities.

The aim was to contribute to local development by means of a new regional organisation, by jointly marketing the region as a good place to live and by lobbying towards regional and national state institutions.

Achievements: Mountain Region Council was founded, and it has supported projects like services for in-migration, forestry industry, decentralised university level education. It has been lobbying for up-keeping railway and airport services and for tax compensations for the long distance transport costs of businesses. The local newspaper has emphasised its role as a voice for the whole of the region.

In the case of the Mountain Region Council in Norway, there was a strong presence of local government, which was worried about the declining numbers of jobs and inhabitants (Rye et al. 2003). In this case the leaders of the peripheral municipalities tried to find new strength by uniting their forces around a new concept of the Mountain Region, and this idea was supported by the local newspaper that took the role of the voice of the Region. Again here, after the initial enthusiasm the process has not proceeded as rapidly as some have expected. However the external networks continue to be active and internal networks have found common tasks in organising the local public services. It could be asked in this case whether the scope of resources mobilised has been narrow. The cooperation has not evolved into new mechanisms of development that go much beyond image marketing and lobbying for public infrastructure funds.

In these three networks the achievement has been more visibility, at least on some scale, but it seems that the scope of resources that were mobilised was rather limited. The lesson from these three networks seems to be that neither public nor private resources alone are enough to go beyond the single projects of building visibility.

Working with Visibility

There are examples of networks where a visibility of a region is an already existing starting point for development efforts. For example in Scotland the entrepreneurs in the area of Skye and Lochalsh have the blessing of the good fame of the region (Árnason et al. 2003). This makes a good basis for many networks of development, some of them quite straightforward marketing networks. Of these one may mention the Skye Network of the Horticulture Association and the new Pebble brand of Skye and Lochalsh Marketing Group. Compared with, for example, the Norwegian and the Irish networks in both of these cases there is a wider scope of actors participating in the process.

Horticulture Association (Scotland)

Actors: In addition to the Horticulture Association, local growers, crofters, locals and in-migrants, Highland Council, Local Enterprise Company (SALE).

The aim is to revitalise and share local skills of gardening and other local agri-food cultivation.

Achievements: Establishment of the Association and advertising its activities: shows, trial plantings, marketing of produce (for example 'Skye Berries'), a weekly stall in Portree town, business advisor contacts for setting up new businesses.

Skye and Lochalsh Marketing Group

SLMG is an umbrella organisation of businesses and other stakeholders developing an overall and more explicit brand for the area.

The aim is to reorganise marketing of local products and services by strengthening the local brand of Skye and Lochalsh.

Achievement: launching the 'pebble' brand in 2003, a new visual logo for local products and services.

Emerging Tensions in the Use of Visibility

There is another network in Scotland that introduces some of the conflicts that have to be solved when intensifying joint marketing. Some of the entrepreneurs participating in the Bed and Breakfast marketing network were not happy with the standard requirements of the tourist board: 'I did object to this grading classification scheme that came in a couple of years ago ... I don't think every B&B should be the same ... And I think they are trying to standardise it too much. Everybody should have this and this in their room and this kind of towels, they even want wallpaper', one told us. On the other hand, some entrepreneurs want to move towards more professional business: 'We wanted to get away from that old image, perception of a B&B being something which was a room, it is more of a business now.'

Bed and Breakfast (B&B) network (Scotland)

Actors: Crofters and enterprises running B&B services, local Tourist Board, and Scottish Tourist Board.

The aim is to intensify marketing of the B&Bs.

Achievements: Improved visibility and wider marketing channels. System for standardisation and classification of services provided by enterprises.

This example of networking brings out one type of conflict that networks have to try to solve when they are building new mechanisms of local development. Here the requirements of product development encounter the earlier local practices of service provision. The new standards are seen to be important for more effective marketing, and they are one step towards defining a network product, the qualities of which are defined by the marketing agent in the middle between the service provider and the client. However, also the local small-scale producers have a point: they may argue that their product is more valuable if it maintains its authentic local features that are a product of its embeddedness in the family life of the host. It is a challenge to the network, how to create a more effective product image that does not lose those local qualities that may have a specific value.

Making Products Out of Place

Our second set of networks consists of those which have moved into developing products that could diversify the local economy of a rural place. Many networks in the case studies have been building products around the image of a place but many have also encountered intricate problems. Should one develop products relying on the earlier products of farms, forests and mines, for example? How can one make a commodity out of the authentic heritage, way of life, beauty of nature and landscape without spoiling it? Who has right to cash in on the value of culture, of landscape, of a pleasant community? How does one organise cooperation and competition around the value of a good reputation?

Although a positive image of the place could be called a common good, meaning that its use does not diminish its apparent value, there is also the possibility that its use can affect its authenticity. Then there is a risk that someone spoils the image, in which case the loss of value is also common. A positive image is difficult to manage, because it is not owned by copyright holders. New users may change the image in the wrong way. In this way images may become fields of battle and negotiation between different camps of users. Creating a brand based on a positive image is one kind of attempt to manage that image, because brands are owned by someone and their user rights are controlled by their owners.

To illustrate these challenges and conflicts, we have chosen three networks whose experience is of specific importance for understanding the mechanisms and issues of local development. The Finnish network around dairy product development in Sotkamo is an example of the reality of competition in a traditional production sector. The case of Dalhalla opera stage in Sweden shows the strength of culture for widening the network resources. In Scotland there are several networks that are good cases of market-based competition build on the reputation of a place. It seems that the name of Skye, the Highlands and Scotland is an inexhaustible source of branding. However, the case of the wind farm proves that this resource can also be an object of contradictory interpretations. The value of the positive image is felt to be threatened, and the process of local decision-making may be stalled by the stalemate of contradictory camps of development.

A Lost Dairy Products Brand

The Finnish case study area, research in which we ourselves carried out, was in the 1980s the regional centre and stronghold of cattle and dairy farming and milk processing in the Kainuu region. In those times the social networks around the dairy were strongly connected to the municipal leadership. The manager of the dairy and the municipal director had also cultivated connections to the national level policy-makers, and in practice these two people and their networks are still very significant for the municipality. The manager of the dairy was a political activist and a personality with a strong will, and she is still regarded as the mother of the particular risk-taking style of local decision-making.

New milk brand network in Sotkamo (Finland)

Actors: Valio Ltd, local dairy, University of Oulu Bio-laboratory in Sotkamo, the municipality.

The aim was to develop and market a new brand of milk products that would safeguard the continuity of the dairy and its jobs.

Achievements: New brand was created and marketed, new products were developed in the newly founded bio-laboratory, and expertise was accumulated in processing products based on milk and also on wild berries and plants. However the brand could not compete on the market controlled by Valio Ltd and the dairy ran into financial difficulties. The dairy was forced to return to the camp of Valio Ltd. (Valio Ltd is owned by 28 cooperative dairies.) Valio Ltd closed the operation of the Sotkamo dairy in 2003.

The intensified competition in food markets has given more impetus to the concentration of the Finnish dairy processing industry. The biggest distributor of milk products is the centre of cooperative dairies, Valio Ltd, which has grown into a large food production conglomerate run by professional managers. Some local dairy cooperatives have been dissatisfied with the growing power of Valio. Small dairies have been closed and the members of the co-ops feel that they have lost their power to make local decisions. In 1993 the leadership of the local dairy in Sotkamo declined the new collaboration agreements offered by Valio. They started to market and distribute milk product under their own trademark 'Aito Maito' (Authentic Milk), and they started product development of their own, such as a family of non-lactose milk products. In the development work they joined forces with some other cooperative dairies that had also declined the terms offered by Valio.

During the 'milk war' the local dairy started losing milk providers as it could not compete with Valio's prices paid to the milk farmers. In September 2000, the dairy again signed an agreement with Valio, whereby the dairy was leased to Valio. Only half a year later, the company announced that the dairy in Sotkamo would be closed in 2003. The closure hit hard, with the two hundred employees losing their jobs. From the farmers' point of view, the closing of the dairy was not necessarily an economic disadvantage: there were other buyers for the milk and even with a better price.

The local milk network started to break down when dairy farmers started to shift to the Valio camp. The fight for the independence of local dairy did not appeal to them as much as better price for their milk. With raw material base crumbling the Kainuu Dairy did not have any future. During the milk war the trust between the local dairy and the Valio Ltd had been spoiled.

The worst thing was this bragging about taking market shares, and the big brother (Valio) did not like that at all. We did have an alternative; we could have worked together with Valio from the beginning, which may have saved some time for our dairy. The costs for marketing efforts of the new brand were enormous, but the dairy administration took that road. Then the producer's prices (for milk) started to go down, and some farmers changed over to the other company. This accelerated the solution. (Tero, male 60 years)

Closing the dairy has also strengthened the suspicions of farmers that Valio Ltd will keep on closing local dairies all over the country and that this highly trusted partner is no longer thinking about the interests of rural producers. 'They did the same trick in Kuopio [another town in eastern Finland]. Now only big is beautiful' (Kalle, male 45 years).

The networks of leading farmers have been an important political force in the municipality. Although other sectors such as tourism have risen and challenged the role of farmers as the backbone of local economy, there are strong continuities of the farmer-based networks and institutions. The strongest among these are the ownership of land properties and the sense of local identity that is connected with farm families' relationship to land and locality. Because of the new growth sectors with bright future visions, however, the earlier partnership of the municipality and milk networks has been replaced by other coalitions.

The attempt to build a new milk brand can be seen as a practical attempt to find continuity for the activities of the milk network. When the development of the new milk brand was underway, a new kind of expertise and research was needed. This was the starting point for a joint project with the University of Oulu, which has a Research and Development Centre in Kajaani. The University founded in Sotkamo (in the very buildings at the dairy) a bio-laboratory that specialised in the research of foodstuffs, particularly milk. The local milk 'know-how' was to be infused with state-of-the-art research expertise. When the closure of the dairy became inevitable, the laboratory specialised in other natural products, such as herbs, wild berries and mushrooms. The establishment of the bio-laboratory was a new form of networking. Some models for this cooperation came from the electronics sector. The bio-laboratory project gave the University an opportunity to strengthen its local impact, and to utilise the financing available in regional development programmes (see also Lehto 2002). The activities of the bio-laboratory were then to be integrated in a planned high-tech centre called Snowpolis, where expertise on health, welfare and sports will be connected with expertise in winter activities. This new research and development centre will be located in Vuokatti, right beside the Vuokatti Sport College.

Opera in an Abandoned Quarry

The Swedish case study relates the story of Dalhalla Opera Stage, which begins, as visitors there are told, with the history of a meteorite falling and forming the rock that was to become the site of a quarry and then an opera stage. The ending of the story is a showcase example of a new cultural economy.

Figure 2.2   Dalhalla concert arena

The process was started by a single person, retired opera singer Margareta Dellefors, who had a summer house in the municipality of Rättvik. She was looking for a place in Sweden where opera performances could be organised in the summer-time. Rättvik happened to have a limestone quarry that was being closed. After seeing the limestone quarry, which was going to be filled with water, Dellefors was convinced that this was the place for the new opera scene. Initially the resistance of the local and regional authorities was like a wall, the idea was disregarded as crazy and unrealistic. But she put her heart and soul, and most importantly, her extensive networks into the project.

Dalhalla Concert Arena (Sweden)

Actors: Retired opera singer and her networks, Dalhalla Friends association, Dalhalla Production Ltd. company, Rättvik municipality, and village associations.

The aim was to establish a concert arena in the former quarry.

Achievements: Former quarry is renovated and it is used as concert arena in summer time. Dalhalla association runs the arena. Local village associations do a lot of contracted voluntary work at the events, earnings from which can be used by the village communities.

The stage was tested in 1993, when about 200 people from the Swedish cultural establishment were invited to a concert. The audience was both amazed and convinced and the local enthusiasts formed a group taking the project forward. The place was given a new name 'Dalhalla' which sounded better than the earlier name 'Draggängarna' (Dragging Meadows) and in 1994 the first concert was opened by the Swedish opera star Birgit Nilsson. During the first season the number of visitors was about 3,000 and in 2001 (including daytime visitors) 112,000.

From our point of view an interesting feature of the success story is the amazing multitude of different kinds of resources that were mobilised into the project. The network of Dalhalla brings together a mix of various kinds of private and public resources, both paid and voluntary work. The municipality owns the land of the quarry, while the non-profit association 'Friends of Dalhalla' owns all the structures that were built to make it into an opera venue. The association has agreement with a company Dalhalla Production Ltd, which runs the business and has a few employees at its office.

The county of Dalarna, and Rättvik as one of its municipalities, is one of Sweden's most attractive tourist areas, especially in summer. Some of the visitors have roots, relatives or friends there. This means that there are plenty of external contacts. This can be seen in the membership register of the Friends of Dalhalla association. Among the special category of members who have paid to get their name on a chair in the arena, the share of persons from outside and particularly from Stockholm area is strikingly high. This category of members has the option of buying their tickets in advance, to make sure that they can get 'their own seat'. Another example of the role of external connections is the fact that the former vice Prime Minister of Sweden Lena Hjelm Wallén has, since 2002, been the chairwoman of the Friends of Dalhalla.

Dalhalla has mobilised large numbers of local people through voluntary work during each season. From the very beginning, there was the idea that the locals should have a stake in the project. They were meant to get some benefits from it, not only the burden of heavy traffic and wear of the roads. The solution was that the five village associations around Dalhalla started to contribute voluntary work. The villagers work as parking guards and concert hosts and hostesses, for example. During every performance about 50 people are needed in running the practical arrangements. These people work voluntarily but their village associations are paid, and they can later invest the money in the village, for example in village halls, a playground or sport facilities. The social meaning of the voluntary work has many aspects: people work for the good of their village, they meet other people in Dalhalla, and they see the performance. However, the matter has also another side: sometimes the commitment may be 'too much' for a limited number of people. If one is committed to the work, it can become hard to manage the volume of engagements. One respondent said:

I am afraid that there are not so many people who are willing to engage to the same extent as us ... Because Dalhalla, it can be really hard in the summer time, it can be too much and too often. (Male representative from a village organisation)

The villages surrounding Dalhalla are small, so it is sometimes hard to find enough people as may be needed for the work. The activists mobilise friends and summer guests, who are interested in music and enjoy working while listening to the concert. While informal networks are a precondition for the project of Dalhalla to run, the activity also has effects on these networks in terms of expansion and innovation.

Results of the Dalhalla project for local development are impressive: more tourists, a longer season and a new symbol of identity. Dalhalla is today one of the main tourist attractions in Rättvik and the whole County of Dalarna. The concert programme has affected the length of the tourist season. It means a lot for different businesses in Rättvik, like hotels, restaurants and shops. Moreover, not least, it has become an important, modern symbol of identity.

The businesses, that is restaurants and shops, talk about the concept of 'Dalhalla-days', that is when there is really a lot of jingling in the cash-box. It is all this which one has worked so long for ... but it took time until the tourist business realised or understood how much Dalhalla means. (Female interviewee at Dalhalla office)

The Dalhalla success story is based on a remarkable network that has managed to connect resources at various levels into the same chain of a unique cultural product. International music stars, national audience of opera lovers, regional and municipal development efforts, and voluntary village work contribute to and benefit from the process.

One feature of the Dalhalla network is that it is very focused around this one cultural product. There are some experiments to widen the cooperation into new kinds of events, for example a country music night during the 'Classic Car Week' in Rättvik. Dalhalla as a product is highly visible and it has established its position in the internal division of labour in Dalarna County, where some municipalities have also created other specialised items of what might be thought of as 'authentic Swedishness', such as the wooden houses around Lake Siljan and the hard breads of Mora and Leksand. Dalhalla's vitalising impacts are limited to the summer season. We do not know of any strong strategic initiatives at the county level, which could intensify and reinforce the connections between these separate products and projects in various municipalities of Dalarna.

Struggling Against and For the Wind Turbines

In the Scottish case study, plans to build a wind farm have involved two camps (or networks) struggling over the use of landscape. Both networks mobilise both local and external resources, and there are contradictory interpretations about what the local interest really is. This is happening in the context of the Scottish and UK governments increasing targets for the proportion of energy that should in the near future come from renewable sources, which includes wind farms.

The network defending the wind farm development consists of the landowner and the crofters of his land, who expect to get future earnings from the international company that will own and run the wind farm. They learned to work together earlier in a fish farming enterprise, although that ran into difficulties. This coalition is allied with an outside company that produces wind energy. The wind farm is planned to generate electricity which will be sold to the UK National Grid by the energy company who will in turn pay the land owner, the tenant crofters and institute a community fund.

The action group objecting to the wind farm proposal consists of people who see local development in terms of tourism and amiable living environment. They argue that the wind farm will compromise the beauty of the landscape and consequently harm tourism in a serious way. The sight of the high wind turbines and the noise of the turbines would spoil the amenity of the township and the economic benefits to the community would be limited, they suggest.

Wind Farm controversy (Scotland)

Actors are divided in two opposite camps:

A coalition supporting the proposal to build wind farm in the north of Skye: multinational energy company, local estate owner and many of the crofters on this estate.

The Skye Windfarm Action Group opposes the proposal: some of those involved in tourist services, some of the new residents, some of the part-time residents, and others.

Results: Although in 2002 the Highland Council's Planning and Development Committee approved the plans, legal challenges to the wind farm continued. Debates for and against the proposal took place in public meetings and in local and national media. In 2007 the Council approved plans for an 18 turbine development.

At a public meeting in 2002, supporters and objectors of the project were given the opportunity to make their case to the Highland Council. The owner of the estate supporting the plan retraced his own roots on Skye: his family goes back at least 500 years on the island and many of his ancestors had been involved in development projects in their time. The opponents of the wind farm, meanwhile, sometimes suggested that the area is still influenced by traditional relationships between the landowner and the crofters. The network that instigated the proposal, objectors say, is not egalitarian and democratic. One interviewee from the opponents said:

> A number of supporters of the Skye Windfarm Action Group are indigenous to this island and have supported us financially but won't put their head above the parapet because they don't want people to know. (Interview, Scottish case study)

The only people who will speak out, opponents continue, are relatively recent newcomers who are not raised in a culture where one should not draw attention to oneself. The Scottish case study shows how the networks make claims about the relationship to land, identity, community and history of the place. There are contradictory claims about the place of crofters. Both sides recognise that crofters represent a kind of legitimate interest in local development and both networks claim that crofters are on their side, if not publicly, at least in their hearts.

This contradiction may offer us some lessons to be thought about. Although the environment, the cultural landscape, and a way of life may be valuable resources and public goods, they may also be contested and their value may be threatened. Secondly, cultural and environmental arguments may be embedded in contradictory projects. The proponents of the wind farm plan refer to their family roots, both land owner and crofter, and their interest in the development and progress of the place. The opponents, in their turn, refer to the continuity of the cultural and landscape values which is valuable part of their own lifestyle and economy. Both sides also fear that the future economic viability of their area is threatened by the claims of the adversary.

Re-organising around the Success

In our first category of networks the challenge was to create visibility – a higher profile – for regions that seemed to be lacking it. Our second category consisted of networks that were creating new products that would be distinguished from others in the market place. In this process the visibility of the place could be an asset. Our third set of networks is again in another kind of position. These cases of networks try to gather local actors and resources around some already successful process, to distribute its benefits to wider networks and to strengthen it further in its success. In these cases the places already have well known success stories about events, enterprises or products. The issue is how to widen the sphere of success to new participants. Here our cases of networks are again of different size but the mechanism of development they are seeking is similar: connecting local resources to an existing successful external connection. One of the cases is food producers' Food Link in the Scottish case study, and the two others are networks in Sweden and Finland that try to transfer local success in sports to other sectors of the local economy.

Food with Local Quality

The area of Skye and Lochalsh in Scotland has a highly visible 'green' image, relating the qualities of the environment there to the freshness and high quality of locally-produced fish, shellfish, lamb, beer, cheese, vegetables, herbs and fruit. The problem is that it can be difficult to buy some of these excellent products locally. As a rule the local supermarkets and shopkeepers get their supplies from central depots on the mainland or from a market in the city of Glasgow (over 200 miles away). Skye and Lochalsh has some potential to increase local production and consumption. This issue has been addressed by local groups, which aim to encourage local food production, marketing and health education. All this is highly dependent on a reliable transport system. Setting up a farmer's market or a box scheme only works if a delivery system is reliable, and associated education only has an effect if healthy food is actually available in local shops.

Food Link project (Scotland)

Actors: Developed by Skye and Lochalsh Enterprise, LEADER II and Soil Association (Food Futures project), Food and Drink officer.

Aim: Localised distribution of fresh local produce. Supporting local producers with networking and marketing opportunities, finding next stage funding.

Achievements: During the three years 1999–2002, a 'Food Link' van service was set up that distributed meat, fish and vegetables from local producers to consumers in Skye and Lochalsh. Van service runs a twice-weekly collection and delivery service from local producers to local outlets including mainly catering, some retail and box scheme customers.

A mobile kitchen unit for cookery demonstrations, and discussions about setting up a Farmers' Market.

A Food Link Van is a solution for local food product transport that was developed by a group of producers across the food spectrum in Skye and Lochalsh (Árnason et al. 2003). This local food link project has been a success and received the national award from the Soil Association as 'Best New Food Initiative' in 2001. The Food Link Van project is part of a national Food Futures programme of the Soil Association. The van is particularly significant because it brought together different sectors of producers, including meat production, fishing and horticulture. It also aided in the development of a local market for these products.

Food Link is in Skye and Lochalsh also called Food Futures. It was funded by the Soil Association, the UK government through its Community Fund, and local organisations. In Skye and Lochalsh, a part-time Food and Drink Officer (in the Local Enterprise Company) became full time for three years to work on the Food Futures project. Although Food Futures was organised and carried out locally, it relied on external funding which inevitably ran out at the end of the allocated time. There was a sense in the interviews for this research that this kind of project may be limited by the difficulties of maintaining the work because it was based on short-term funding, necessitating frequent repeat applications. Interviewees argued that a longer term funding would make projects like this more viable. Conflicts may also exist between such new initiatives, which arrive with large but short-term funding, and the on-going, and potentially more socially embedded efforts of more local organisations or groups. Developing food and drink in Skye and Lochalsh has been a long-standing development strategy in the area.

The organisers of the Food Futures were in a difficult situation: without a financially viable van there can be no increase in local food production and without more produce the van has no long-term viability. The Food Link van was not expected to be self-supporting for some time (Scottish Executive 2003).

The Food Link project is an attempt to increase the value of relatively limited production of local quality foods and it aims at the higher end of the market. It is based on innovation of food distribution (meat, fish and vegetables) and networking of producers of different production sectors and suppliers (hotels, restaurants). The local action is connected firmly with an idea created at national level and supported by funding from UK government and Lottery fund, and partly financed by from regional level (via the then Scottish Executive) and local funding from SALE (the local development enterprise). These external connections give it some 'top down' qualities that may be a source of conflicts and that may make the maintenance of the results more difficult.

Gathering Around the Ice Hockey Ring

Leksand is a small industrial town in the county of Dalarna in Sweden. It has been proud of its the ice-hockey team IAL (Ice Hockey Association of Leksand) that has played in the Premium League since the early 1950s. Two events made the town and the county rethink the role of ice hockey in regional development. The first setback came when the one and only big sponsor of the team, a big timber company Stora Kopparberg wanted to withdraw. The second one was the team losing its place in the Premier League. These two shocks led to reorganisation of the relationships between the ice hockey team and the region's communities.

Sport and business network in Leksand (Sweden)

Actors: Ice-hockey club, private sponsors, and Leksand municipality.

The aim is to contribute to local development by utilising a strong ice-hockey image of the locality and creating new cooperation between sport and business for enhancing the local economy.

Achievements: Traditional combinations of voluntary work in sport events and other sport and business activities. Innovative aspects: new business meetings in sport events outside of the region and utilising the networks of big enterprise directors for attracting business to the region.

Changes in the sponsorship network meant a thorough reorganisation of the team's external connections. For decades the team had been a heroic small town team supported mainly by one big company. The withdrawal of the timber company from this contract was a historical change and the construction of the sponsorship on a new county-wide basis was a huge accomplishment. During the next season (2000–2001), the number of main sponsors increased from one to eleven, and the sponsoring revenues were more than doubled. All the sponsors have their base or origin in the county of Dalarna. The group of sponsors was given the name 'The Gold Team'.

The second shock came in March 2001. After having played in the Premier League for 50 years, the elite team of Leksand Ice-Hockey Association lost its place there. After this the core group of the association and its new Gold Team started a spectacular rescue drive 'Dalauppropet' (the Dala-Appeal). Also in this campaign the Ice-Hockey Association of Leksand reached out to the whole county of Dalarna, not only the town of Leksand. The chair of the Ice-Hockey Association was a businessman with extensive connections, and he used them.

> The Dala-Appeal. Yes, that was also ... a bit unique in Sweden. I still had my old contacts at the newspapers, so we went around and asked them first. And we got media; well we got 150 whole pages for free, so that we could go out there and market us. Everyone joined in, even a marketing company. It is unique, I think. So we worked with fantastic marketing, the largest which has occurred in Swedish sports. It was also important, that when we fell out of the Premier League, people felt: 'Oh, they drive hard anyway'. So I think, that we fell out, it was maybe good for us. (Chairman of the board of Leksand Ice-hockey Association)

The campaign also referred to the long-standing relations of trust between the sport club and the community:

> To collect money is nothing new for an association in a small town like Leksand. In this way, the artificially frozen ice-rink was built in 1956, and the ice-hall in 1965. (Homepage of IAL, 9 September 2002)

At the end of the summer 2001, IAL had gained 2,000 additional members and 700 new sponsoring firms who had bought the so-called 'advertisement kit' of the campaign. Nationally famous sports-people and artists signed up for 'their' team in a wide media-campaign. With the help of the additional revenues, some new players were recruited and in April 2002 the team regained its place in the Premier League.

The transformation of the ice hockey networks of Leksand and Dalarna did not happen without side effects. The identity of a town team was changed into a county team. The network enjoyed a common experience of succeeding in winning back the place in top League, which may have helped to foster a mutual sense of community among the businesspeople of the county. These contacts were not limited to the county because the team started to organise entrepreneurs' meetings at the games in Stockholm. Connections to prominent business people are used to attract companies to the region and to create new contacts. The events and games of IAL have become a meeting place, the value of which may be reflected in the business results.

As the networks of the ice hockey team are changing, someone has to manage the rising new tensions, such as that between business and voluntary work. Both the leaders of the team and its critics have recognised that the ice hockey team is turning into even more into a business. It is a seventh largest employer in Leksand. Its results are calculated in yearly turnover (around 40 million SEK). Some of its activities were organised in 1995 into a limited company running a restaurant, conference services, a souvenir shop, and some marketing activities. The organisation of the Ice Hockey Association itself has become more business-like: it has a managing director, who is a former elite player. The chair of the board is a prominent local businessman and the members of the board are men (only men) representing firms of tourism, banking, insurance and the information sector.

However, the success of the IAL is based on organising sport activities, and it requires large amounts of voluntary participation and voluntary work. About 1,000 people are voluntarily engaged in various ways. Some are unpaid coaches of junior teams. Many adults, mainly parents of boys playing ice hockey, are engaged around the teams: maintaining equipment, working in kiosks, selling lottery tickets. Others, not only parents, work voluntarily at the matches as parking guides, door attendants, in the kiosks, etc.

The elite team and the junior ice hockey have different budgets and the junior ice hockey is run on a non-profit basis. It is very important to inform the parents in detail about the budget. Sometimes parents of the boys playing wrongly assume that their voluntary work (for example selling lottery tickets and working in the kiosks at the matches) contributes to the costs of the elite players.

Our researchers, Tillberg Mattsson and Stenbacka, ask whether there is a danger that the commercialisation of sports, which on the one hand can have positive effects on the local economy, may on the other hand have negative effects on the building of social capital. For example, the representative of the Business Association in Leksand talks about how things have changed. Suddenly, the scouts, who have always helped at events like the celebration of Walpurgis Night (the night before May Day), expect to be paid to do so. The woman who is organising the youngsters' cup games in Leksand sees no drawbacks in that but she has noticed that it becomes more difficult to get people committed in voluntary work:

> I think you have to reverse a bit more in this; there should maybe not be so very much money in it ... It should become a fellowship to realise something. (Female volunteer engaged in the junior ice-hockey)

Nowadays, it is not only members of ice hockey networks who recognise that ice hockey plays a decisive role in local development. This opinion is prevalent also among municipal and business representatives.

> Leksand is a bit specific in that; we usually compare it to a millipede, where the municipality is the backbone and the Ice Hockey Association the head, and entrepreneurs are the thousand legs. (Female representative of the Business Association of Leksand)

The case of ice hockey in Leksand brings out the tensions that are created when private enterprise-like activity is run with the support of voluntary work. Getting money provides for new resources but at the same time it can create new divisions.

Connecting Skis, Spas and Baseball

The Finnish case study of the municipality of Sotkamo similarly relates a story of local reorganisation of networks around success in sports – in this case baseball and skiing. Sotkamo is also a small municipality with a visible top League sport team. In Sotkamo we may witness a further integration of sports and the rising of a new economic sector, tourism.

In Sotkamo the decline of farming and forestry as a primary source of livelihood has been compensated by the development of new sectors of activity, which intertwine previously separate developments in winter sports, the Holiday Club and Spa Hotel (summer tourism) and baseball. The outcome has involved turning the Vuokatti landscape into a common brand name for tourism and sports, and a symbol of Sotkamo's future.

Vuokatti network (Finland)

Actors: Municipality of Sotkamo, local sport associations of Sotkamon Jymy and JymyPesis, local tourism enterprises, Sport College.

Figure 2.3   Ski tube in Vuokatti

The aim was to engage in local development through new cooperation in marketing sport and tourism.

Achievements: Coalition of sport and tourism marketing, brand of Vuokatti

Vuokatti Sport College, in addition to being an international centre of ski training, becomes a centre of new tourism developments. The attraction of Vuokatti as a winter sport centre is based on the combination of many kinds of winter sports: the slalom slopes and ski lifts, 100 km of cross-country skiing tracks, ski jumping hills, facilities for biathlon training and competitions and an ice hockey hall.

The founding of the Katinkulta Spa Hotel and Holiday Club created a large-scale tourism enterprise that runs on a year-round basis as part of an international network of holiday clubs. Nowadays Katinkulta Holiday Club is able to market itself as 'the most versatile holiday and congress oasis in the Nordic Countries'. Supporting this claim are its developed sport facilities (including tennis, golf, bowling, the spa, gymnastic hall, and many outdoor tracks) and the network of small entrepreneurs offering additional services that range from moped or snow scooter safaris to overnight trips to the wilderness or a visit to organic farm serving local food.

The rise of Katinkulta changed the networks of tourism marketing. The municipality-driven political culture of 'one man – one vote' was replaced with business approach of 'one mark – one vote'. The municipality became a silent partner, while the marketing network was led by the larger operators, Katinkulta, Vuokatti Slopes Company, and the Sport College. Smaller enterprises could buy into the marketing campaign by contributing to the costs.

Katinkulta has changed the general thinking about tourism. In earlier years tourism was a way for farmers and the Sport College to get some extra income. Nowadays tourism is regarded as serious business with good future prospects. The new networks of Vuokatti marketing have brought new foresight and planning in the tourism marketing. Together the key partners sell both winter and summer experiences to the travellers. Jointly they can boast to offer both summer (Tropical Spa Hotel) and winter (Vuokatti and ski tube) seasons all round the year, which gives them an advantage if compared with, for instance, holiday resorts in Lapland. The marketing circle began to plan new events to fill the quiet months of the year, thus increasing the use rate of their accommodation and restaurant capacity.

The rise of the Sotkamo baseball club Sotkamon Jymy to the position of the most successful baseball team in Finland has created huge visibility. The success has created local enthusiasm, which has mobilised local people and enterprises.

> The championship game, for example in 1995 between the favourite team Oulun Lippo and Sotkamon Jymy, have been struggles between traditional centre and periphery or between the two counties. (Kolamo 1998, 59)

> These moments are enjoyable in many ways. Firstly, our audience has the experience of the game, the ecstasy of winning. The impact goes much deeper than just me or the team being successful. This whole environment, the spectators, those listening to the radio, all the people of Kainuu Region and Sotkamo get the feeling of having succeeded. This game has proved that masters may come even from such a small place like this. (Former star player, Väisänen 1997, 106)

The success of the baseball club can be traced back to earlier successes. The club had won the Finnish championship once before, in 1963. After that there was a period when the club dropped out of the national top League. At the end of the 1970s a long march towards new successes was started. The fruits of the long-term junior work were harvested in 1990, when the team won again the Finnish championship after 27 years. With the success of the baseball team, an important shift took place in the tourism marketing networks of Vuokatti. The joint marketing circle of tourism enterprises decided to buy marketing services from the baseball team, which could use its media contacts and visibility as a resource.

Figure 2.4   Vuokatti logo in tourism and sports

The baseball team enterprise SuperJymy Ltd has focused its marketing efforts around the image of Vuokatti, which has become the common brand of local tourism marketing. Now sports, tourism and municipality are all using the same marketing image, Vuokatti Hill. The name of the municipality has been put in the background and the brand of Vuokatti is brought forwards in all fronts. This strategy is different from that of regional marketing.

> In earlier times they have tried to market the idea of the Kainuu Region, but people are not interested in coming to Kainuu for forests [Kainuu has been seen as a region of forests]. For those living in Helsinki there are plenty of forests much nearer. We are now making Vuokatti into a brand, and in principle others may join us if they will. We are not going to join Kainuu regional marketing. (Pekka, male, 37 years.)

In the context of the Nordic countries the role of local government is central in local development. It has the mandate for land use planning, it has tax incomes of its own (although pressures to cut down public expenses are severe), and it is the organiser and provider of the welfare services. In addition it offers a local forum for participation and political struggles, as its decision-making body is elected by general vote.

During the recent decades the municipality of Sotkamo has changed from a local welfare organisation into a development agency. While in the 1970s Sotkamo municipality was in the forefront of building modern health centres and social services, it is now an innovator of image construction and the development of tourism. Sotkamo municipality has been one of the main actors in organising various development projects. The Vuokatti development has received a lot of resources from outside the region but to get them one also has to have local funding. The municipality is the most important source of local funding (for example matching funding in EU projects). This is a reason why local decisions and struggles about development strategies take place in the arena of municipal politics.

The leaders of the Sotkamo municipality were not very happy with the plan of regional welfare strategy of Kainuu Region. According to the municipal executive board 'it defines Kainuu to be a sunset county' of the elderly. Sotkamo wants to attract people of active age who enjoy working and living in a holiday land of sports. The same emphasis can be found in the marketing of Sotkamo both for residents and for tourists.

> In our marketing we always use Vuokatti and its violet colour. Our target group is over 35 year olds who can afford to drive Volvo or Mercedes and who are living around Helsinki or Oulu. This assumed group of clients has funds and we want to collect their spare money. (Director of Sotkamo municipality)

The Sotkamo success story has the support of the many active groups in Sotkamo. Those who are satisfied with the results like to mention the growing numbers in tourism, increasing investments and good publicity. However, not everyone is happy about the rise of new tourism and recreation business. Some critics say that Vuokatti Hill area is being turned into a 'tourist and sport ghetto'. In their opinion, too much of the municipal effort and investments are channelled to tourism and to Vuokatti village, while services in other parts of municipality are declining.

The sport associations and events have brought together different groups, and given them positive experiences of working together and being successful together. However, the coalition of sports and tourism has a polarising effect on local solidarity. The sport life of Sotkamo has been divided into two camps, the national level sports consisting of skiing and baseball, and the more local and grass-roots voluntary work of smaller sport clubs. In addition, the elite of sport and tourism seems to ignore the voices of cultural and village activists. The fields of sport and tourism are seen to be dominated by male values. There are female voices in particular demanding more resources for rural development, for public services, and for culture (arts and heritage activities). The activists in cultural groups feel that the significance of culture for a good quality of life is not recognised by the decision-makers, and neither is its long-term significance for the development of tourism and general attractiveness of the area.

Conclusions

In this chapter we have reviewed the networks for development that have been reported in the case studies. These observations should be put in the framework of the three aspects of networking that were defined in the beginning of this chapter.

Networks that are in the beginning of building visibility for a new regional entity seem to be based mainly on local resources, with modest external resources in the form of a project or a contribution of a development enterprise. One may observe that the responsibility for local development may be taken by different kinds of actors, depending on the role of the state. In Italy there is a connection between local enterprises and a regional public programme, in

Ireland local enterprises or groups and a semi-public development enterprise, and in Norway the leaders of local municipalities.

The second group of networks, which are in the process of creating new products utilising the good visibility of the place, have as a rule a larger scope of resources ranging from national (or even international) to regional and local levels. Often they are networks that follow the value-chain of the product from the producer to the consumer. Our cases demonstrate both risks and possibilities in new products. In the case of the Sotkamo's new milk product brand, the products themselves were advanced and of good quality, but the marketing efforts were too demanding for the available resources of the network. The network lost out in the competition with bigger market players. During the struggle the old relationships of trust were spoiled and the role of milk networks in local politics declined. Some players in the milk network moved to other networks. In the case of conflict over wind power in the Scottish case, the networks were organised in two opposing camps with contradictory interests in the value and the use of landscape.

The third category of networks is already dealing with some kind of success and is trying to reorganise local networks, to join local forces to strengthen and utilise this success. Again the amount and scope of resources may vary. The Scottish case of Food Link consists of local food producers and a nationwide organisation promoting organic food, and regional development agencies. The national and regional interests bring new resources to local actors who want to use the positive image of the place (including good natural conditions) to improve the distribution of their product to local consumers, and simultaneously to increase local demand. Here the concept of a local product of good quality is a resource, and the network has produced a practical innovation of transporting the products of the network. The external resources are brought in as a project with a limited time span, which seems to create a risk for the continuity of new activities.

In the case of Leksand Ice Hockey Association in Sweden, the original scope of resources were local – the sport organisation with its voluntary activists and a sponsor which was a large timber company with a local office. The case study describes how the resources of the network were reorganised after two traumatic events (withdrawal of the sponsor and relegation from the Premier League). The network of sponsoring resources was widened and extended from the town of Leksand to the county of Dalarna. At the same time the nature of the network was changed to open new kinds of resources (improved connections) for the participating enterprises. The results have been a success, the hockey team is back in the Premier League, but new tensions have arisen between the business-like functioning of the hockey team and the mobilisation of voluntary activists, which is recognised to be vital for success.

The case of Vuokatti network in Finland describes a successful process of widening the scope of resources mobilised in a network for development. The winter and summer tourism has been joined into network that is utilising the high visibility and marketing skills of the successful baseball team. In addition, the municipality is deeply involved and committed to the new strategy of development, which has however created new tensions. Villages and cultural activists feel that the strategy is concentrating too many resources on tourism and sport entertainment. The municipality has also distanced itself from the regional strategies of development, emphasising its own priorities and target groups of development policies.

Dimensions of Social Capital

How do we interpret our observations about networks and their activities in terms of social capital? The aspects of networking that are used as a frame of observations can be linked with specific and separate features of social capital. *The scope of resources* that the network opens for its participants is directly connected with the *extensiveness of the social capital*. Access to resources of various kinds – knowledge and cultural resources as well as property and funding – may be opened via networks. The extensiveness of social capital is not only about the number of participants in the network nor the total volume of resources available, it is also, and may be in the first place, about the multitude of the kinds of resources available. In this sense, from the viewpoint of single actors, social capital is contributing to the activities when it brings access to such resources that otherwise were not accessible. In some cases this may be an issue of quantity of resources but in many cases it is an issue of additional qualities of resources.

The other features of networking analysed in this chapter are related with the way of acting and decision-making in networks. The decision-making capability is linked to a dimension of social capital that we may call the *thickness of social capital*. It refers to the idea that social capital builds not only bridges between different resource holders, but also enables them to agree on concerted and focused action. Not only are resources combined together in creative ways to make new tools of development, but these tools are also used in a decisive way. This sharpness of action is very much based on the depth of the mutual trust and understanding of the network partners, an acceptance of the necessity and legitimacy of making decisions and practical actions without delay. It creates support for those partners who are ready to act for the benefits of the network. It also allows for taking risks and accepting consequences of mistakes and failures without tearing apart the basic fabric of the network.

However, just trusting others to act and giving them loyal support may lead to rigid path dependencies and expensive mistakes, of which we use as an example the failed attempt to win markets for a new milk product brand in Sotkamo. As important as loyalties inside the network (based on bonding social capital) is the ability to process lessons and to experiment with new actions. Social capital could be used to support collective learning processes in and of the network. We have tried several words that could express the connection of the networked learning to social capital. It has to do with the agility of the network, its capacity to react to changes in the environment. It has also something to do with the sensitivity to weak signals, openness to various forms of communication, an atmosphere where disagreements and critical sentiments are encouraged to come forward into the open. It presupposes such levels of mutual trust that one may express differing opinions without a fear of losing one's face or credibility. It accepts mistakes as a normal part of a process where mistakes become one part of learning together. This kind of network is tolerant of conflicts and may have many ways of solving and dealing with them. Such trust may grow out of long experiences of successes and failures. In terms of the qualities of social capital this collective learning capability could be connected with the *vividness of social capital*.

We can condense our general conclusions based on the cases into three statements:

1   the multitude of network resources relates to the extensiveness of social capital, which widens the possibilities that network members have in accessing and combining resources to be used for common activities;
2   the general trust and mutual understanding of network members, called here the thickness of social capital, enables decision-making and effective action in spite of conflicts and different opinions;
3   the relaxed and open forms of communications, a vivid motion of social capital in the network, supports collective learning processes, which in turn enables flexibility and creativity in social practices.

If we combine these three dimensions of social capital, we see that social capital has many faces, and it may have both positive and negative impacts on local development. At one end of the scale there is social capital that is narrow, thin and stagnated and at the other end of the scale there is social capital that is extensive, thick and vivid. Extensiveness brings in resources, thickness creates trust and enables effective and risk-taking decisions and vividness gives flexibility and helps to learn from experiences.

Skills of Constructing Social Capital

Some general lessons may be drawn from comparing the networks for local development. First is the notion of virtuous circle of social capital. According to this idea, communities with strong voluntary activities that have an autonomous relationship to administrative structures are able utilise initiatives and resources coming from the state for strengthening their own horizontal networks (Warner 2001). Conversely, in those communities where local networks are dependant on the authorities, new external resources tend to enforce the existing social relations and contradictions.

The second general observation is that successful networks for development manage to put together factors that seem to be contradictory. We may list several opposites, the balancing of which seems to be related with the above mentioned virtuous circle of social capital: external and local resources, commercial and voluntary activities, official and unofficial organisations, common image (identity) and commercial brands, short-term flexibility and long-term strategy.

Thirdly, it is important that networks have a central node, or network of networks, where all the creativity and effectiveness and conflict resolution takes place. It is evident that this network of networks may be maintained by different institutional structures, such as local government or local development enterprises.

Chapter 3
Social Capital in Rural Areas: Public Goods and Public Services

Claudio Cecchi[1]

Introduction

In this chapter social capital is considered to be part of the total assets belonging to a community, together with physical capital and human capital. Therefore capital, as composed of different parts and owned in a community (by single members or by the community as a whole) contributes to the development level reached by the community itself. In economics and in other social sciences, the social dimension of capital cannot be neglected in the explanation of the performance of a community and the outcomes that the community can expect.

While there is a wide consensus about the necessity to increase awareness of the social dimension in the notion of capital, a clear and common definition of social capital is still missing: each discipline and sub-discipline adopts, sometimes, a specific definition that someone else frequently disputes.[2] As a consequence, it is necessary here to show clearly what definition of social

1   This chapter is one result of the activities carried out by the Italian team within RESTRIM. My gratitude goes to all members of the team. Manuela Menichetti collected the largest part of the questionnaires, both in first and second round, moreover she has carried out some important interviews. Andrea Micocci has written some basic material on social capital and has analysed the questionnaires. The entire design of the research has been decided together with Elisabetta Basile who is the scientific coordinator of the team. I cannot mention here all the people we have met during the fieldwork in Maremma, but mainly to them goes my gratitude for the understanding of the functioning of the economy and of the society in this marginal rural area. Finally, Fabio Sabatini (University of Roma 'La Sapienza'), and many members of other RESTRIM teams have sent comments on a previous version of this chapter; to them goes my gratitude together with apologies in case I have not been able to take into account of their suggestion. The usual disclaimer applies, because the responsibility for any mistake and/or omission in this text is mine alone.

2   In economics, for example, Kenneth Arrow has discussed the correctness of using the term 'capital' with reference to social relations or institution. Arrow's critique is that some pre-requisites are missing, like the willing motivation of the actors who invest in building of social capital (Arrow 2000, 7).

capital I am assuming and to give reasons for the choice. In this chapter I will assume the very broad definition that Grootaert and van Bastelaer present. The authors define social capital as 'the institutions, relationships, attitudes, and values that govern interactions among people and contribute to economic and social development' (Grootaert and van Bastelaer 2002, 2).

According to this definition, social capital is the result of the use of resources – that might have been used in a different way – whose benefits can influence the performance of the community for a long period of time. Other forms of capital share the same characteristics, but there are significant differences. Physical capital is made of produced or natural material goods. Human capital, on the other hand, being immaterial, is represented by skills and other abilities that are embedded in any single member of the community. However, immaterial goods that are part of the assets of the community as a whole can be said to constitute social capital. For these reasons, the outcomes of social capital assume the nature of public goods. The characteristics that distinguish the outcomes of social capital as public goods are: i. the outcomes of social capital can be 'used' by each member of the community (non-excludability); and ii. the use of social capital by any member of the community does not exclude its use by other members (non-rivalry).

Being a public good, each outcome of social capital distributes many different types of benefit to all members of the community. These benefits assume the form of services that each member of the community can obtain. In order to analyse the role of social capital in development processes, we have to distinguish its use – i.e. the services that social capital provides – from the subjects and the social organisation that can potentially provide the services. In academic literature, the benefits from social capital in the production process are represented by the services that the social capital provides to the members of the community who share it (Robison, Schmid and Siles 2002, 9–10).[3] In particular, three categories of services are supplied by social capital: i. economic services that do not pass through the market and are subject to market rules; ii. social services provided by means of personal relationships; iii. validation services that help people (and economic actors) to be considered and recognised as a trustworthy members of the community.

In relation to the subjects that embody the social capital, there are two different dimensions that give different perspectives through which social capital can be defined and scrutinised: the *forms* and the *scope* of social capital (Grootaert and van Bastelaer 2002).

Different *forms* of social capital refer to different degrees of cohesion in a community. A community that has a low level of cohesion can be described

---

3   The authors actually consider four types of services; the fourth being 'services of encouragement, moral support'. Here, I consider this fourth type as part of the previous one (validation) because the latter also logically implies, in my opinion, different forms of support to the members of the community.

and analysed only by reference to the 'structural' feature of social capital: that is, with reference to the body of formal rules that govern local behaviour. The structural dimension of social capital refers explicitly to institutions, networks of relationship and formal codes that govern economic and social actions. To explain the behaviour of a very cohesive community, reference has to be given to the 'cognitive' dimension of social capital. This dimension refers to more subjective and intangible elements that contribute to the governance of social and economic actions. In particular, we refer to values, behavioural rules, and norms that define trust at a community level. The particular form that social capital assumes within a community contributes to the way in which the community faces the needs of its members and gives collective answers to their explicit and/or undisclosed requests.

Different *scopes* of social capital focus the attention on the amplitude of the subject that defines, creates and changes social capital. Starting from the micro level, we observe individuals who have chosen to share their interests, their needs and their actions on the basis of a common aim or of a common problem. At the meso (middle) level, many different communities decide to cooperate or, at least, to interact on the basis of their contiguity that can be defined at a geographical dimension, or at a social or economic dimension. Finally, the macro level of social capital is observed when many different micro actors and meso level actors define ways that help in creating and controlling general components of social capital. Such social capital belongs to, and can be used by, each member of the community as a single individual or as a community at a lower level.

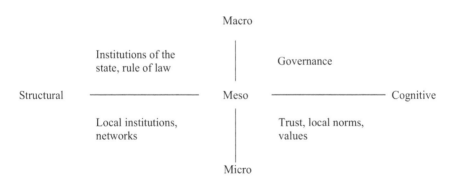

Figure 3.1   The forms and scope of social capital

Source: Grootaert and van Bastelaer.

Figure 3.1 synthesises and shows the two dimensions of social capital, by means of examples that distinguish the meaning of the four different quadrants. Moving from top-left towards bottom-right, we found that in the beginning, social capital describes a community governed from above with a very low level of 'community life' and, at the end, a very exclusive community that works mainly through specific behaviours, with a high degree of isolation.

One can easily note that, when speaking of social capital, it is always necessary to make reference to a community. This reference represents a problem that has to be solved both in theoretical and practical terms. In other words, the researcher should state, on the one hand, the theoretical framework within which the analysis is developed and, on the other hand, what kind of community is investigated.

Here we refer to a local community: the community that constitutes the group of persons who belong to a *local economic system*. Social sciences have indeed shown that the local dimension of development is the only one that can be considered in order to understand any process of growth, of development and of change. The analysis of the national and regional performance is realised by means of the sum of the results shown by the local systems that the nation, or the region, includes.[4] Few variables that describe growth and financial phenomena, can be considered only at an aggregate national level. Exceptions are the international rate of exchange, the rate of interest and some parts of the state's budget. At local level, these variables must be considered as exogenous. All other aggregate variables have always a relevant local component.

Our research concentrates on rural local systems which are considered to be remote. In relation to the dimensions that have been taken into account for the study of social capital, we can observe here that the analysis has focused on the central and lower part of Figure 3.1. We assume that there is some sense of belonging to the community amongst the citizens who live in the areas we have studied, and that it is worth scrutinising the entire range of variation of social capital, including the structural and the cognitive dimensions. The survey that we carried out mainly focused on the analysis of local institutions and networks in Southern Tuscany – the *Colline interne dell'Albegna e del Fiora* and the *Amiata grossetano* local systems. These areas are two out of five local economic systems included within the project *Maremma: The Rural District of Europe*, enacted by the administration of the Province of Grosseto through the Rural Regional Plan of the Tuscany Region. Moreover, we have also explored the system of local norms, values and trust shared by the population.

In order to perceive both sides of social capital – the structural and the cognitive character – our research has analysed many different types of indicators. The most important aspect that I wish to highlight here concerns the level of the services that the local community can use, in terms of quantity and quality. Furthermore, some indicator of performance is considered in order to measure the results of the development strategies that the community has taken part in.

---

4   This is the well-known Becattini's thesis, based on the 'partial equilibrium' approach that he derives from Alfred Marshall (Becattini 1999).

These choices, in terms of indicators, allow comparisons between the case study areas. Our research theme concentrates on public and private services as indicators that depend on the presence of social capital, and that contribute to the building of local processes of development. Our hypothesis concerns the relationship between social capital and development. As shown in the literature on social capital, there are important relationships between the quantity and quality of social capital that characterises a community and the level and type of development achieved by the community itself (Sabatini 2003). More precisely, the purpose here is to explore how far the action of local institutions in supporting development projects and strategies is influenced by the contribution of the social capital that characterises the community. This contribution may have a positive or a negative influence. It is likely to be positive when the community is able to play an active role, for example by proposing new projects and actions. It can act negatively if, for example, a single group tries to seize all the advantages produced by development programmes. The reason why some communities show a more sensitive attitude towards collective actions than others rests on the experience that the members have based in past results of public action. This is particularly true with reference to services. I suggest that there is a correspondence between the situation in which a need of a citizen is satisfied through a collective action or service, and the same citizen trusting and invoking collective action as a means to enhance development strategies.

The assumption that sustains our hypothesis relates to the relationships between social capital and development. First of all we assume that a higher level of social capital gives a community a stronger capability for planning new initiatives. Then we assume that a low level of social capital implies that the initiatives are implemented only if they give private individual benefits. Thirdly we assume a community that has high a level of social capital carries out initiatives that, amongst other things, makes the social capital itself increase and improve.

The following section describes outcomes of social capital have already been considered as public goods in literature, and outlines the logical path that links the concept of social capital with the goods represented by 'services'. This link is examined through the observation that the outcome of social capital is a public good and that many public goods are represented by some kind of service.

The subsequent section – the Maremma backwardness in terms of services – concentrates on the relationship between the supply of *public* services (education, health and quality of life) and the level of social capital that characterises a community. Here we start from a presentation of the Italian welfare system and refer to the study area for some evidence based on historical developments. With reference to the present situation, we observe that the marginality, in territorial and social terms, does not diminish even if some signals of economic growth appear. In order to sustain this statement, the evidence comes from the analysis of statistical information about the availability of services and about the income in the area; and also information given by a sample of the population by means a questionnaire.

We then explore some relevant differences between the case study areas used in this book. More precisely, we look at what the different communities have in common and what differs between them in terms of social capital and in terms of development strategies. We examine the differences between the study areas in terms of availability of services and in terms of networks, and finally, the processes of reproduction of the social capital that characterises each of the study areas.

We make the case that there is a causal link between the availability of services and development. Here the availability of services is considered in historical and in present terms. Services are considered in both the public and the private spheres, and development is considered with reference to the strategies that each community plans and its actual performance. The argument is made that there is a relationship between the marginality of a community and the availability of public goods which, at the same time, contributes to the understanding of the process of social capital building in which a community is in a position to develop. In terms of the relationship between social capital and development, we show that there is both a theoretical framework and evidence to justify the principle that social capital has a major role in determining the performance of development strategies and actions.

Social Capital and Services

Social capital is frequently considered to be a public good. For example Woolcock (2001) draws on Robert Putnam in adopting this definition:

> Social capital is generally perceived to be a private and public good (Putnam 2000) because, through its creation as a by-product of social relations, it benefits both the creator and bystander. It is a classic public good because of its non-exclusivity – its benefits cannot be restricted and hence are available to all members of a community indiscriminately.

This definition emphasises the possibility of all members of a community gaining from *the use* of social capital, which is considered here to be a 'by-product of social relations'. The following quotation by the World Bank emphasises instead the conditions that enhance the production of social capital, suggesting that the cooperation between institutions can create a fertile field for development.

> The broadest and most encompassing view of social capital includes the social and political environment that shapes social structure and enables norms to develop. This analysis extends the importance of social capital to the most formalized institutional

relationships and structures, such as government, the political regime, the rule of law, the court system, and civil and political liberties. This view not only accounts for the virtues and vices of social capital, and the importance of forging ties within and across communities, but recognizes that the capacity of various social groups to act in their interest depends crucially on the support (or lack thereof) that they receive from the state as well as the private sector. Similarly, the state depends on social stability and widespread popular support. In short, economic and social development thrives when representatives of the state, the corporate sector, and civil society create forums in and through which they can identify and pursue common goals. (World Bank, n.d.)

This quotation introduces explicitly the state as a subject that actively participates in the process of creating social capital. Here, the state sustains private action because it contributes to the creation of the social and political environment. The important role of the state in this action is justified here through the *public good* character of the outcomes of social capital. How, then, does the state create or support social capital?

Taking social capital as an example of the basis for the production of public goods, one can easily observe that no single firm will be interested in producing a public good that is non-excludable and non-rivalled in consumption. On the contrary, it will be easy to find people who over-consume the public good or who have no hindrance to deplete it, if its use gives them some satisfaction, utility or cost reduction in some economic activity.[5]

A broader interpretation of public goods can be taken to include goods that are not public goods *per se* but are goods that entitle people to exercise a public good. For example, railways can be considered as public goods 'if one assumes that the right to freedom of movement means that one citizen cannot be excluded by virtue of an appropriation by another citizen' (Bianchi 1998, 111).[6] Public goods are therefore a consequence of a 'natural' or an 'institutional' market failure, which prevent the supply meeting the needs of the population and of firms.

The shortage of a public good can therefore be addressed by means of action implemented by the state or by some group. The difference between these cases refers to the resources used to produce the public good and to the people who are entitled to use the newly produced public good. The state and, in general, public administration, choose to supply a public good using public

---

5   Economists define these situations as 'externalities', which can be positive or negative. The externalities are positive, if each consumer increases his/her welfare as a consequence of the presence of social capital: are negative if the presence of social capital decreases the welfare. The same applies to firms in terms of costs and/or returns.

6   In this case, railways can be considered also as meritorious goods. The term 'meritorious goods' describes private goods that are institutionally transformed into public goods for normative reasons. Another real example is represented by electric energy, which from 1962 to 1989 was supplied by a state monopoly and considered in Italy an example of a meritorious good.

resources coming from direct or indirect taxation. The state decides to produce it by means of state owned structures or to buy it from a private enterprise. The state decides whether or not to make consumers pay a 'contribution' – not a price – for the use of the public good.

Some groups, such as private associations, formal and informal networks, can choose to supplement the shortage of a public good by means of a 'private' supply. The group chooses the resources to be used for the production of the public good: from the pocket of a single member or from the pockets of all members. The group decides how to produce the public good: by means of structures owned by the group, or buying it from a private producer. The group decides whether it supplies a proper public good (with non-excludability) or it supplies a quasi-public good (reserving the use to the members of the group), which represents a substitute for the lacking public good. In summary, social capital represents for a community a stock that can provide public goods. Public goods can be increased or consumed by means of the action and of the choices of the state, of groups and of individuals.

Public Goods in Rural Areas

It is necessary to emphasise the fact that we are dealing with rural areas. Therefore, we must take into account some specificity that distinguishes this type of territory. With reference to the evolution of the Italian economy during recent decades, rural and agricultural local systems make an important contribution to the overall path of development because rural territory covers more than a half of the national surface area and contain more than 40 per cent of the Italian population.

The rurality that emerges in recent times appears as very different from that of pre-industrial Italy. Basile and Cecchi (2001) have shown that this difference is as relevant at the theoretical level as it is on the basis of the evidence coming from the Italian countryside. Post-industrial rurality is characterised by the complete integration of rural areas within the contemporary economic and social organisation of the capitalistic world. This organisation is different from other local systems because production is differentiated between and within sectors. The dispersal of economic activities generates a rarefied social fabric that is, under many circumstances, the antithesis of the industrial district. Furthermore, the presence of a completely integrated agriculture, amongst other activities, represents a characteristic element that contributes to the singularity of the economic system and the qualification of the social system.

Many factors underlie the change just described, but one of the most important is considered to be the reversal of the direction of the resource flow between countryside and town. In mid 1970s, the flow of resources between countryside to town in western Europe and North America changed in character. Since then, historians, sociologists and economists have tried to build models for the interpretation and prediction of the direction of resource flow, noting that simple models of flow from country to city no longer hold. None has reached a unequivocal conclusion, but many have observed that different performances can be observed in different local and regional contexts.

By the end of the 1990s, economists were able to describe two different outcomes for the countryside. The first relates to areas where the modernisation of agriculture has resulted in agricultural specialisation. Alternatively, where the modernisation process has failed, a large part of local resources remain under-employed and available for a different use. In this second situation, 'modern' rurality emerges because rural resources have attracted urban ones.

Modern rurality is then characterised by a rarefied social fabric, economic differentiation and the presence of agriculture. This is the consequence of the interaction between urban and rural resources. Modern rurality is frequently considered to be a positive situation, because it represents a new vitality for declining social organisations. Nevertheless, a limiting factor exists regarding the process of development and of growth for rural areas. This limit is represented by the potential shortage of services, needed for industrial growth, for resident families and communities.

In other words, it is true that urban areas expel human and financial resources, and it is true that rural areas are suitable for new activities. But the availability of services generates new competitive advantages for each different type of territory.[7] Traditional services, financial services and 'modern' services, when suitable for the rural local economic system, can sustain growth and development, or can constrain them when not integrated in the economic and social organisation.

We have analysed the role of services within the development processes, which involve rural and agricultural local systems, in three different directions. The first direction concerns the definition of specific characters that distinguish rural and agricultural development processes. This considers the connections between the supply of services and the functioning of the development process. In the case of services, we refer to two different origins of services: public and private. While the former supplies public goods that start and sustain growth processes, the latter emerges on the one hand as a consequence of the structural transformation of the economy and on the other as a manifestation of the needs of firms and people within the local system.

The second direction concerns the construction of a typology of services suitable for the support of a sectorally differentiated and locally integrated

---

7   The local systems literature examines the consequences of post-industrial transition. In this literature, global transformation, together with suitability of local resources to face the transformation itself, explains the functioning of the model of local production system's and local social system's change. This model was originally born within the industrial district analysis (Becattini 1987), but it has been extended to the more general category of the 'local system' (Becattini 1989). The industrial genesis of the local system concept makes difficult the use of the model for the analysis of rural territories. Nevertheless, Cecchi (2002) has shown that 'rural' and 'agricultural' local systems represent true ideal-types of local systems.

economic growth process. This research direction points to a specific typology for rural areas. In fact, in this kind of territory, the economic and social organisation emerges as the result of rural needs but also the starting point for a new prevalence of market forces. Rural needs are different from urban needs in that the community lives in a different context: markets for rural firms appear usually to be limited and therefore an enterprising rural firm should search for new opportunities outside the area.

This typology of public services must be analysed with particular attention in rural areas, because the characters of the local community entrap their supply in a dangerous trade-off. High production costs, partly due to the limited dimension of the supply and the limited number of tax contributors, and also to the absence of private supply because of the limited number of consumers, make it difficult to find the right dimension of public supply.

The third research direction concerns the identification of the subjects of governance of rural development processes. The research has concentrated on the multidimensional composition of the development process. The components are: the planning procedure, the management and monitoring practice and, finally, the control and the accountability of actors. The opportunity for this analysis comes from the observation that rural development processes are frequently considered to be endogenously produced; but many authors also believe that they are generally governed by external institutions. Sometimes, one can observe a formal institution that governs the process by means of formal rules and decides the resource use; this is the case for a public administration empowered by consensus. Alternatively, one can observe informal institutions that drive the growth process by means of substantial control of the local resource use; in this case there is private empowerment of resources.

Within this theoretical framework, public goods are strictly linked to public action and to the action of civil society. Our interest in public goods is confined to the local dimension; this assumption implies that we are not interested in public goods such as defence, which is usually considered to be an archetypal public good. Our focus is mainly on services within the study areas that constitute the way in which the state contributes to the building up of public goods. In other words, we have tried to identify the kind of action with which the state, by means of the local administration, increases the welfare of the population and makes new economic initiatives appear attractive. But we are also interested in understanding the ability of the local community to face and supplement any lack of state action. Therefore, we looked for plans implemented by the civil society – networks, associations, and so on – that produce or develop local public goods.

We focus on three types of services, which refer to different dimensions of the public good component of social capital: education, health and quality of life. Specifically, we focus on the creation of those services; in other words, we highlight the subject that has the power to supply the service.

The Maremma Backwardness in Terms of Services

The analysis of the creation of social capital by means of combined state and private action is here considered within a framework that has two different aspects. The first concerns the process of change that has influenced the study area during the last decades. Our case study area of Maremma (Figure 3.2), like Italy as the whole, has experienced during the last half of the twentieth century an important process of change that has been acted out differently in different parts of the country. The second aspect concerns the welfare system that has characterised the post-war period of rapid and intense economic growth: during this period, the welfare system has paid attention mainly to urban areas and to the industrial labour force, sustaining only marginally the rural population.

Figure 3.2   The village of Sorano in Maremma, Italy

The Welfare State and Services

The role of the state in providing services for the population and for economic initiatives is linked to the welfare state system that prevails in a country. Italian history is characterised by the strong presence of a central administration that has always maintained a dominant position in terms of planning actions: the welfare system, the education system and the other public services have been centrally governed in every single detail, even if regional administrations have been involved in the management and financing process.

We now mainly focus on the welfare system, which actually shapes all decision processes in the provision of public services. Furthermore, we consider some consequences of the prevailing characteristics of the welfare system for the way in which other services are managed. According to Himllert (2003), the Italian welfare system can be understood to be *familistic*, and thus significantly different from the welfare state systems prevailing in Northern Europe. Ferrera notes:

> Italy has a mixed model of welfare. Income maintenance benefits are provided by occupationally fragmented social insurance funds, while health care is provided by a universal national health service. A variegated, largely decentralised system of social assistance provides cash benefits and basic social services to various categories of needy people, such as poor elderly or the disabled. (Ferrara 1998, 1)

The reason for describing it as *familistic* lies in the fact that families and family networks represent the main source of informal assistance among citizens. The main characteristics of this system are that benefits are strongly linked to (and dependent on) labour market relationships, and that the larger part of the expenditure goes towards maintaining the pension system. As a consequence, there is a bias in the way in which the welfare state is financed. The pension scheme, which provides more than the two-thirds of social benefits and includes also social security pensions, is mainly financed by workers and firms. Only one-third of the social benefits are financed through taxes, such as for the National Health Service that is financed now also by means of contributions paid by people who use some type of health service.

The decade that shows the most impressive growth of social expenditure in Italy is the 1980s, when the social benefits grew from 19.4 per cent to 24.1 per cent of GDP, with an annual percentage growth in real terms of 5.5 per cent (Ferrara 1998, 5). After this period, Italy has suffered the same crisis that has hit all industrialised countries; this crisis has fostered a structural reform of the state expenditure that, nevertheless, has not yet changed the weight of the welfare system expenditure.

Within this framework, the social benefits – measured as share of GDP – are at a lower level (24.3 per cent) than the European Union average (26.2 per cent). The peculiarity of the Italian welfare system gives rise to other important structural ways of working. While a largely decentralised system of social assistance seems to prevail, the pension system, absorbing the largest part of the benefits, and the health service, largely run at national level, represent the most important parts of the welfare state system. It is evident that the management of the pensions and health service is not decentralised at all. In fact, with reference to the remaining part of the welfare system, we should speak more of fragmentation than decentralisation.

The Italian education system has its own peculiarities but it is important to highlight the fact that, whilst the structures (buildings and other materials) are owned and managed by local administrations, the rules that govern the supply of education services are defined centrally. This refers mainly to the quantity and quality of teachers and the teacher/pupil ratios.

Under these circumstances, the capabilities of localities to adequately supply education according to the needs of the population is very low. This is particularly true with reference to rural areas, where the population is usually more dispersed that in rural areas. According to official statistical data, the area – in common with the province as a whole – has a quite low rate of pupils per teacher. This indicator also hides the reality represented by a large number of 'special'[8] classes, where this indicator is not useful. From our survey, we have observed a very high number of special classes in the area: in the *Amiata grossetano* LES, something less than one third of the classes are of this type and, in *Colline interne*, half the classes are in this category. Moreover, with reference to high schools, we can observe another peculiar situation. In numerical terms, the situation appears to be reasonable because ten schools provide places for approximately 1,700 youngsters; but the geographical distribution of these schools suggests that there is actually no choice for the students because distance become a very strong constraint. Therefore, unless one chooses to commute daily or to move to another locality, each pupil can only attend the nearest school.

It is worth mentioning that there is no university nearer than 100 km from the centre of the area. Moreover, very few workshop activities take place in the Colline interne LES. Therefore, the limited supply of higher education and the absence of jobs push youngsters (and sometimes their families) towards migrating from the area.

A similar situation is found in the health service. We must distinguish between the structures and the availability of services. On the one hand, in terms of beds within the two hospitals of the area, 4 beds per 1000 inhabitants is a quite high number if compared with urban areas. The total number of 130 beds shows that the type of assistance given within these structures is limited to ordinary practices and some emergency. People in the area are used to travelling to Grosseto, Siena or even Roma or Florence for the most serious treatments; and maternity care is also not present in these local structures. We should mention here that ambulances are quite numerous in the area: there are 30; moreover there is an emergency helicopter service based on the hospital of Grosseto. In relation to the decentralised supply of health assistance, we must argue that the supply is very poor. For a total population of 31,000 citizens only 170 hours of assistance by a medical doctor is available in eight centres; the average distance from these centres is 14 km.

---

8   Special classes in Italy are those made up of groups of children with two teachers who include children of different ages or with handicaps.

The services we identify as relating to quality of life (such as job centres, tourist offices and police stations) are present at a standard level in the area. The reason for this is mainly that these services are centrally managed directly by the government, through its ministerial offices. In this situation, it is quite surprising to note the complete absence of private initiatives. In fact, whilst private delivery services and private security activities – like couriers and private guards – represent very important and growing initiatives in urban areas, in our study area they are not developed at all.

Furthermore, in our fieldwork we came across some protests about all types of public services, from different perspectives. One person deplored the sewage systems that was said to leak and cause pollution, and who also stated that the waste disposal system was discontinuous; another complained about failures in the electrical energy distribution and in the water supply system; and, lastly, many lamented that the mobile phone system covered only a small part of the area.

From the answers given to questionnaires that have been submitted to a sample of the population in the area, and from reports on other public meetings in the area, we can suggest that the population in general does not perceive there to be a lack of good public services. More that 40 per cent of the respondents considered the quality and quantity of all the public services in the area to be 'adequate', while only 15 per cent consider the level of all the public services to be 'inadequate'. The most common opinion in relation to the kind of service that should be increased and improved relates to transport, both in terms of routes and public lines. The transport service only involves buses and is now in fact completely private, with only small subsidies given to the bus companies. Amongst the services that are considered to be inadequate, education occupies the first position, and is described as such by 25 per cent of the respondents.

The Lack of Collective Private Initiative

The combination of the structural transformation and the changing welfare system has shaped the initiatives undertaken by the population in the area. The change that has affected the economic structure of the area has removed a large number of traditional jobs: the mining sector closed down in the 1970s, and agriculture has witnessed a massive introduction of technical change that has caused an astonishing reduction in labour requirement. Furthermore, during the 1950s the state concentrated mainly on agrarian reform, and since then welfare state prevailing in Italy has largely failed to meet the needs of a population who lived in an area with a declining level of opportunities. People in Maremma, as with other Italian agricultural and marginal areas, obtained public support only through agricultural policies, and this was a lower welfare support than industrial workers or people in towns. Under these conditions, private initiatives were mainly oriented towards profit generating activities. Migration to growing industrial areas represented the most popular alternative for a population lacking in resources.

Few collective initiatives were successful in the area and those that were have been undertaken by cooperatives that received strong support from national regulation. The greater part of the cooperatives in the area are still linked to the agrarian reform and to the special support programme that followed the mining crisis. The cooperative system appears to be the result more of the pressures coming from the 'reform agency' and political action than of a common will. When, at the end of the 1990s, the cooperative system faced a crisis as a consequence of the decreasing state support, the most important change in the area was the privatisation of a significant number of such cooperatives (mainly in the food production). Solutions to economic problems appear to be guided by private individual initiatives, whilst collective action is confined to non-economic activities or to activities that do not satisfy basic needs.

The lack of public services has been addressed more by means of exits from the community than by the search for other people who shared the same need. As an alternative, private individual choices have supplemented the shortage that has been due to the lack of state action. The population continues to migrate from the area, even as there is also a significant amount of immigrants. These immigrants apparently require a lower level of public services or, because they are often wealthy, they can afford to buy private services from mainly outside the area. The recent concern for the environment in rural areas could promote collective actions, but instead the prevailing trend is for many forms of privatisation of the environmental dimension to the community. The interviews we have carried out in the area show that, in the opinion of the people, development actions improve certain private earnings more than the quality of life of the community as a whole.

A Comparative Perspective

In order to make some comparison between the case study areas, it is now necessary to follow the same path as that undertaken in the analysis of the creation of social capital in Southern Tuscany. We focus first on the different welfare systems that are present in the north European countries we are dealing with.

Welfare Models

First of all, it is necessary to observe that the welfare systems of Norway, Sweden and Finland share the social democratic or Northern model (Hillmert 2003). They show the greatest and most significant differences in comparison to the familistic Italian and southern welfare system. Citizenship is based on a completely different assumption, producing two types of communities. While in southern countries, the family and family networks are assumed to be the source for the provision of assistance, which is then supplied in an informal way, in northern countries the family, and particularly each individual within it, is considered to be the target of the assistance system supplied by the state. In southern countries the state plays a role confined to the provision of a limited social security (concentrated mainly on citizens who participate in the labour market) and on the definition of rules that govern the action of local administrations. In northern countries the state rather builds the social infrastructure and provides services by means of a broad network of public structures (Lehto and Oksa 2002a).[9]

Another characteristic of the social-democratic welfare system is the importance of local administration: in northern countries, the organisation of services is chiefly undertaken by municipalities even if their degree of freedom varies in relation to different kinds of services and to different national contexts. For example, whilst in Finland 'the main responsibility for producing social, health, and educational services has belonged to the municipalities' (ibid.), in Norway one can say that 'municipalities are characterised by a "negative freedom limitation". That is, they have to execute the tasks they are required to by law, besides taking on whatever tasks they want to' (Rye and Winge 2002a). In Sweden, municipalities have power to collect taxes on the basis of local strategies (Jonsson, Rydén and Tillberg 2002). Even if there are significant differences between countries that use the same Nordic welfare model, it is important to emphasise the most important character: the role of the state is mainly to *produce* services, while in southern countries the role of the state is to *redistribute* the revenue by means of taxes and expenditure.

The second comparison concerns Britain and Ireland: both these countries are considered to employ a *liberal* model of welfare state (Hillmert 2003). Although the liberal system leaves the state with an apparently limited role, we can notice that social benefits from welfare state amount in Britain to 25.8 per cent of the GDP.[10] This situation is very different from Ireland where, after a long period of reductions, the share of social benefits has reached the level of 13.4 per cent of GDP – the lowest level in European Union. Even though Britain has witnessed a long period of privatisation, the role and the commitment of the state is still very important in planning and financing the provision of public services and sustaining the welfare system.[11] The process of privatisation has

---

9   There are some significant differences between Nordic countries; for example it important to stress that in Norway, 'basically, the welfare state provides certain free services to all citizens without regard to income, but few of the services are unconditional' (Rye and Winge 2002).

10   A value that is higher than in Italy, for example. In year 2000, according to Eurostat (2003), the share of social benefits on GDP in the European Union as a whole is 26.2 per cent. The same index for the other case study countries is: Italy 24.3 per cent, Ireland 13.4 per cent, Finland 24.4 per cent and Sweden 31.7 per cent.

11   It is also important to note that there is considerable concern over levels of public services in Skye. According to the Scottish case study, decline in public services affects remote areas even more strongly due to the additional difficulties in accessing them.

accompanied a de-centralisation of services, and today there is 'a fragmentation of responsibility to a host of non-elected bodies from central state, private sector and civil society, necessitating partnerships to pursue "area-based integration"' (Nightingale 2002). In Ireland, the liberal model 'combines modest universal schemes and extensive means tested assistance. The state intervenes on a discretionary basis on the principle of need. It intervenes to protect people who are unable to protect themselves, either because they belong to a high-risk group or because they are not attached to the labour market. Benefits are paid at a flat rate and are financed out of general taxation' (Kinlen 2002a).

While some recent changes have influenced this model of welfare, the degree of participation of local populations in the process of managing public services is nevertheless still high and linked to the planning and managing structures of the local administration. There are differences between Britain and Ireland that are worth stressing. they concern mainly the 'confidence' that the citizen has in the action of the state: whilst in Britain people are used to the support of the community as an organised body, in Ireland the trust of the citizen lies in the capacity of the state to perceive specific needs by means of its decentralised structures. In other words, the two systems differ in the role played by the official networks of relations.

In the areas that use the liberal or the social democratic welfare model, people who have been interviewed by the research teams for these case studies do not generally complain about the level of public services, they only complain about the change that follows the crisis of the late 1980s and 1990s. This change is represented by a diffused reduction of the quantity and quality of public services. Nevertheless, we can easily observe that the quantity and quality of services in the rural areas we are dealing with are often high, if compared with the Italian case.

Public Services

In order to compare the level of public services in the study areas, it is important to stress the fact that the density of the population in the areas highly varies amongst countries. These differences partly explain also the differences in the indicators of the quality of public services.

It is of course not only the density of each area as a whole which is important, but also the distribution of the population within the territory. In the Italian case, for example, the population is scattered in ten municipalities, and each of the includes not less that three villages, each of them including a large number of *case sparse* (isolated houses, which are usually farm buildings); therefore, the population is widely spread through the territory. The northern areas present a low population density that usually corresponds to a population concentrated in villages. This circumstance makes the provision of services easier, on the one hand, and more expensive, on the other. It is easier because a small community

Table 3.1   Population density in the case study areas

	Inhabitants/km^2
Sotkamo (Finland)	4.2
Skye and Lochalsh (Scotland)	4.4
The Lake District (Ireland)	8.6
Maremma (Italy)	36.0
Mountain Region (Norway)	3.2
Leksand and Rättvik (Sweden)	10.0

requires little assistance by the state and is able to identify in a better way the needs of its members; it is more expensive because if some level of welfare is granted to every citizen, in order to fulfil this commitment, the state must devote resources even if the number of beneficiaries is low.

In terms of indexes describing the quantity and quality of services, northern countries do not show a significant difference with the Italian case. The number of school rooms (primary and secondary levels) is proportionately comparable to the results shown in the Italian case, and almost the same applies to hospitals and to some other services linked to the 'quality of life'. Nevertheless, there are significant exceptions. The first one relates to the area of Leksand and Rättvik (Sweden), where municipalities make all possible efforts in order to maintain public services (at least education, health, and child and elderly care) at the high level reached in the past (including state provision of broadband internet for some years). Another exception is the Kainuu region, where great importance is attached to education and elderly care: this area is characterised by a wide range of educational opportunities, including the Kajaani Polytechnic that hosts 1500 students and is less than 40 km from Sotkamo (the centre of the Finnish study area). The last important case is represented by the Norwegian Mountain region, where there are two hospitals (95 beds) for a population of 25,000 and an area of 14,000 km^2 (more than ten times the Italian area).

The major difference in comparison with the Italian case comes from the fact that all other study areas show a significant presence of private initiatives. These initiatives have three possible characteristics: the first is the production of private services comes from a specific state choice, as it is in case of the post offices that have been given over to private businesses during the 1980s and 1990s in many European countries. The second case refers to private services acquired by the public administration, for example transport for pupils from home to school. Thirdly, private initiatives deal with services that the state, or the local administration, does not provide, such as some kinds of specific professional training courses. By screening the questionnaires that were submitted as part of this research we can make a tentative evaluation as to the perception of quality in public services in the case study areas, bearing in mind that samples were not intended to be representative of the population, but merely broadly descriptive.

The areas in countries that adopt a social democratic welfare state model show a high level of satisfaction about public services. For instance, in the Norwegian Mountain Region, public services are top of the ranking of reasons for living in the area (Rye and Winge 2002b). In Leksand-Rättvik (Sweden) 'public services – including health care and care of children and elderly – is in total ranked highest, though women rank it slightly higher than men' (Tillberg Mattsson and Stenbacka 2003). In Sotkamo (Finland) 47 per cent of the respondents consider the quality of public services in the area to be one of the three best reasons for considering it as 'a good place for you to live and earn your living' (Lehto and Oksa 2002b).

A quite different situation concerns Ireland and Britain that share the liberal welfare system. Referring to the Lake District, the Irish team writes that '[we] asked the respondents what factors in the area they are not content with and why. Half of the respondents elaborated on this question on detail, showing some of their concerns for the area. These included issues such as: lack of services, particularly in relation to healthcare' (Kinlen 2002b). With reference to Skye and Lochalsh, only 6 out of 62 respondents to the questionnaire of our Scottish researchers considered services to be important (Lee and Árnason 2003).

Finally, in Italy, where the familistic welfare system prevails, only 6 per cent of the respondents consider public services as a good reason for staying in the area, and more than one-third of the respondents consider 'lack of structures and infrastructure, and bad public services' to be the most significant thing they do not like in Maremma (Micocci 2002).

Welfare State, Public Services and Social Capital

In order to complete the comparative analysis, our aim is to show how communities participate in the process of building new and structured initiatives as part of a network or association. In other words, we link the degree of participation in common local life to the process of building social capital.

In this research, the study of social capital has focused on the analysis of networks and the participation of members of the local community in networks. Attention has been paid to networks that have been built on the basis of public action and the functioning of public administration; but the greatest attention has been paid to networks that are 'privately' managed and those that people build as a 'place' to share common interests and concerns. Moreover, the interviews and other fieldwork have concentrated on the analysis of participation in networks that relate to development initiatives and cooperation with planning authorities.

In Sotkamo (Finland), all the respondents belong at least to one (formal) association and, amongst them, there is a group of people who play an active role at least in four networks, named by Lehto and Oksa (2002, 31) as *super-active*.

The role of networks in the planning process is very important and is part of the democratic government system that characterises this country. Respondents to questionnaires consider development actions implemented during the recent years to be successful more because of the degree of their participation than necessarily for the economic result of the actions themselves (see Chapter 2).

The degree of network participation in the Norwegian Mountain Region appears to be very high because all the interviewees are member of at least one network or group; and amongst them almost half play an active role as part of the management committee of a network or group. The degree of information and participation in public action in the area is quite surprising.

> The informants were asked to name what agencies are the most important actors in the area regarding economic development. Many actors from different sectors are mentioned: business actors, the municipalities, the counties, the state, NIRDF, local associations, local and central politicians etc. Other informants name concrete persons or enterprises. None of these agencies/actors obtain distinctly more references than others. However, the general impression is that the various state or semi-state agencies are perceived to play the most important role. (Rye and Winge 2002b, 12)

In Leksand and Rättvik (Sweden), networks play a very important role because they represent the basis for the construction of trust amongst the citizens towards public administration. Moreover, networks that show a high level of cooperation between citizens and between networks themselves can be seen as 'a natural continuity and expressions of the social activities and networks that are, and always have, existed in the region' (Tillberg Mattsson and Stenbacka 2003, 35). In Sweden the participation in networks and the voluntary work of the members show the old tradition of 'village democracies' in action (ibid., 15). The role of networking in building plans and other development initiatives is emphasised by the Swedish team in this research, who note that success derives from the secondary position played by key actors and from the capacity to also involve the large number of entrepreneurs who are not embedded in networks of civil society.

> [R]egarding both the question of services and the stimulation of local business development, non-key development actors expect and trust the municipal politics and authorities to be responsible. A strong vertical trust could thus be seen as hindering innovative ideas concerning the realms of services and entrepreneurship to emanate from 'below', from civil society. (Tillberg Mattsson and Stenbacka 2003, 50)

In Skye and Lochalsh (Scotland), whilst the number of associations and networks is lower than in the other northern areas, the degree of participation in the activities of those groups is high: 75 per cent of the sample have between two and six memberships. More than 53 per cent of the respondents asserted that they play an active role as a member of the management committee of the association, and almost the same percentage stated that they have been actively involved in development initiatives. This high degree of participation is associated with a diffused success of public initiatives that have also involved citizens and associations.

> It should be mentioned here that it is regional agencies that are charged with promoting economic development in the Skye and Lochalsh region and so high scores for the local agencies seem to suggest that respondents are content with the promotion of economic development locally. At the same time, the results suggest that respondents are less than content with those areas of development that are perceived to be the responsibility of the national governments. (Lee and Árnason 2003, 28)

A similar situation was observed in the Lake District (Ireland). The degree of participation in networks and groups does not appear to be a relevant aspect; in fact, the people who interviewed were not able to describe any activities in detail when asked about any roles of their associations in development actions.[12] Moreover, the link between association and the public administration appears as vague, and the associations do not see any consequence between their action and development (Kinlen 2002b, 18). For example: respondents were asked about to whom they would turn if they were setting up a business; 'some people did not reply or said they did not know ... Of those who did reply many mentioned specific individuals working in agencies and their individual qualities' (ibid., 10). The Irish case highlights how people appear not to be able to describe their role in development 'despite mentioning their involvement in various community and voluntary associations in the area' (ibid.).

The case of Maremma (Italy) is significantly different. The degree of participation in associations and networks is lower than in any other case study area under consideration here. Only one third of the sample gave any answer to the question related to the type and number of associations belonged to. Moreover, few people say they have an active role in the functioning of the group. In relation to the counterpart of public action for development, it appears that only a few types of association play a role. This is the case of political parties, which are obviously strongly linked to representatives in the elected local bodies, the workers' unions, and the associations of entrepreneurs and farmers. The evidence given by respondents is the long list of associations mentioned as

---

12   Kinlen (2002b, 10) says: 'The groups in which the respondents to the questionnaire appear to be most active are charity/voluntary groups, community councils and sports and cultural groups. Whilst some stated that they are active in such groups, when asked about their role in development they answered that they were not involved and therefore did not necessarily see any link between their group involvement and "development".'

relevant for the planning activities: the people in the sample do not identify a limited group of specific subjects that link the administration with citizens.

The major consequence of the weakness of the links between planners and citizens is the low level of awareness of what development actually means for a citizen and about the ways of planning development actions common in the other case studies. In this sense it is quite surprising that, in the ranking of measures that have been appreciated in the context of previous programmes that should be improved by means of future actions, the first position in occupied by 'roads', while the second is a generic reference to tourism (Micocci 2002, 11). The reasons for satisfaction or dissatisfaction about previous programmes are diverse. While the most positive answers are linked to personal earnings, the negative answers claim the 'distance' between citizens and planners and the fact that planners 'just do their minimal duty' (Micocci 2002, 18).

In summary, we can observe that there is a relationship between the quantity and quality of formal and informal networks of citizens and of associations, and the capacity of the public administration to address the needs of the population or, at least, to make the population aware of public actions. The consequence of this is that networks of relations increase and improve their way of functioning by means of public support. For this reason, we must emphasise the necessity of paying attention to social capital in all its components in order to understand the process of change in rural areas.

The Role of Services for the Restructuring of Rural Areas

We can relate these findings to the more general theme of public policies for rural development and their funding. On the one hand, the concern comes from the reduction of public expenditure, or at least the necessity of reducing public expenditure as a consequence of the commitment of the European Union members' governments to maintain the so-called Maastricht parameters within the agreed range of variation. This has major effects on the welfare state and on the provision of public services that should be granted to the population. On the other hand, the eastern enlargement of the European Union raises problems linked to the risk of another change of emphasis in the Common Agricultural Policy (CAP) in favour of the support of agriculture instead of economic differentiation in rural areas. This is because agriculture is still a very important activity from many points of view in these countries.

The consequences of this twofold recent threat concern the entire functioning of the CAP, from the growing importance of private initiatives – the supremacy of market forces – to the strategies to be implemented for rural development support. The regional and social equilibrium, which was the basis of the Treaty of Rome since the 1960s, has been affected by the failure of the mechanisms introduced with the CAP tools to support the agricultural sector and by the great change that enveloped the countryside at the end of the twentieth century.

The CAP of the 1990s promised to give support only to farmers who were becoming true entrepreneurs by being attentive to market forces. The same policy has also promised support to rural communities that were able to plan development actions by means of internal forces in the so-called endogenous development strategies.

This change of the aims of the EU's policy was based on the assumption that social and regional problems might be solved by private action and that the only role of the government is to supplement resources to those who lack them. This assumption is wrong. In other words, it is wrong to assume that farmers always behave in order to maximise their profits and it is wrong to assume that the local communities have always had a strategy that allows them to fill the gaps that separate them from the richest areas.

Focusing on the social capital that characterises a number of rural communities in marginal areas, we have shown that the capacity to define and implement development strategies depends on the availability of social capital. This in turn depends on the history of the community in terms of the results of public action. Moreover, the success of the public action depends on the type and on the level of services that the state provides to the local population. The effectiveness of rural policies does not depend on the target of the public funds transfers in terms of farmers or other components of the community, nor on the specific strategy that each single community is able to plan. It rather depends on the capacity of public action to create the conditions that make any action, and the resources that are necessary for its implementation, efficient and productive.

The social capital that characterises a community significantly influences the capacity of the governance of the community to produce new initiatives, its capacity to manage the implementation of initiatives, and the capacity to link top-down actions suggested by the central government to local private personal or collective actions. The quantity and the quality of social capital depends on the past history of public action in terms of the supply of public services. From the analysis of the link between the needs of the population and the ways in which each citizen can meet those needs, it emerges that in those localities where the 'normal' means of provision comes from public services, people trust collective action as a means of solving both private and common problems. In other words, communities with a good level of public services – as in localities belonging to countries that use the social-democratic welfare state model – have a high level of awareness of the role of collective action and of the significance of public services as public goods.

One may expect that, as a consequence of the reduction of the supply of services by the state, the network of local relations can create the possibility for the substitution of local collective action for the state action. This causal relation appears to be stronger when the level of participation of citizens in the decision process is higher, in other words, the level of democracy. This tradition of democratic participation in the decision making process, or the degree of

Comparing Rural Development

coincidence between local government and local governance, explains the reason why development programmes can find the answers to local needs by means of local resources and of external public funding.

These observations have consequences for the design of European rural development policy. The consequences can be listed in three different groups:

1   In relation to the target of the transfer of common funds, where the aim is economic. It matters less who the target of the funds is than that this target is an active part of the social fabric of the community. Public funding should respond to needs expressed by the community as a whole even if the beneficiary of the transfer is a single person. This is a good way to ensure public funds for development instead of them becoming private gains. It also means that agricultural policy, for example, could direct more funding to farmers if are more an integrated part of the community instead of just the agricultural market. Considering targets to be an 'active part of a community' raises the problem of social exclusion, however. In this case, other forms of funding must be planned which have a social rather than just economic aim. In other words, marginality needs to be addressed by a specific policy and should not be confused with economic weakness.

2   In relation to the type of action that the policy should support by means of public funding, where the aim of the policy relates to regional disparities. Marginality is a condition that makes public expenditure less efficient than in more integrated areas. Poverty of social capital is at the same time a cause and a result of marginality. It is a cause because communities that have received a low level of attention by the state and, as a consequence, have 'consumed' their social capital, are not able to properly use public funds. It is a consequence in that marginal communities, because of their isolation, are not able to collectively use the state resources and to make the social capital increase and improve. Therefore, public action that aims for development should be directed to the type of projects that increase social capital.

3   In relation to specific aims of rural development policy, where development means an improvement of the quality of life of the citizens. The history of state action, in terms of public services, explains the quality and quantity of social capital that characterise a locality. It has also been shown that in a community where public services have been supplied properly, social capital is well developed and the population trusts the collective action. Therefore, a policy that takes for granted the capacity of a community to plan its own future ignores the fact that those abilities depend on the level of trustworthiness in collective action and, moreover, that the trustworthiness depends on the past public action. It must be stated that each action programme of rural development has to be combined with a programme of growth of public services that sets

the conditions – the quality of life – of the local population to a similar level of an urban integrated population and to conditions similar to the ones of other communities that enjoy of a better quality of life in other areas of the European Union.

In order to implement a productive strategy of development in marginal areas, and many rural areas are of course marginal, a policy that enhances development should present certain characteristics. First of all, it must consider the quantity and the quality of public services and of all public goods. It must enable the public administration to meet the basic needs of the population in order to create a climate of trust towards collective actions. Such a policy should also rank the projects that the community has planned according to their capacity to address common requests. These projects should receive a high degree of priority for the funding, in order to respond to needs that are perceived as collective by the members of the community. Lastly, projects that have a strong economic impact should be considered for their capacity to increase the earning power of local resources, which might then be used for alternative activities in other areas (as Lehto and Oksa suggest in the previous chapter). The aim of this is to increase the degree of integration of the owners of the resources within the local community.

Here, then, we have presented a rural development policy that is rather different to that currently used by the European Union. We argue that social capital is at the root of the process of building and implementing local collective actions, and that trust in collective action is a distinguishing character of social capital. Moreover, the analysis of communities has shown that trust in collective action depends on the role played by the state in terms of answers to individual and collective needs. Finally, our analysis has shown that the main role of the state at the local level is to supply public services. In other words, a lack of public services generates a community that does not trust collective action, and, as a consequence, willingness to build social capital is very weak.

When a community is poor in social capital, the state should attempt to supply a better level of public services in order to create a climate of trust in collective action. It is only within this climate that the community will invest in processes of building social capital, and it is only on the basis of high levels of social capital that the local community can plan and implement effective rural policies. Within this framework, investment in building social capital becomes a priority in public action.

Chapter 4

Gendered Social Capital: Exploring the Relations between Civil Society and the Labour Market

Susanne Stenbacka and Karin Tillberg Mattsson[1]

Introduction

Earlier research has been able to establish a connection at the community level between lively participation in voluntary organisations and a flourishing economy. This has been explained in terms of dense, informal networks of civic engagement fostering norms of reciprocity and trust – described as social capital – which in turn facilitate cooperation and successful economic development. Engagement in voluntary work is thus commonly considered to be positive in influencing community building and economic life in favourable ways. However, as the link between civic engagement and economic development has been described only at an aggregated level, the actual relation between the two has not been highlighted. In addition, the connection on the *individual* level, that is the relationship between engagement in the voluntary sector and position on the labour market or in local business, has not been sufficiently explored.

Focusing on the relationship between voluntary engagement and economic development only at an aggregated level also means that possible differences in the quality of social capital might be disregarded. There could be structural differences, but there may also be variation in the processes by which power is wielded and transferred within networks. By way of applying a gender perspective and focusing at the level of the person, this chapter contributes to understanding the relationship between voluntary work and economic development. The theoretical framework is centred around the concept of social capital and especially the gender dimensions of social capital. What can we

1   The authors would like to thank all the people in Leksand and Rättvik in Sweden who have given their time to our project. Many thanks to Inger Jonsson, Uppsala University, for her work during the first year of the RESTRIM project. She contributed with valuable ideas and was engaged in the preparation and preliminary analysis of the questionnaire study. We would also like to thank the members of the research teams in the other countries who have shared research findings, information and ideas.

learn about gender dimensions of the associational life and of the relationship between voluntary engagement and paid work?

The empirical findings originate from a study in the two municipalities of Leksand and Rättvik in the region of Dalarna, Sweden (see Chapter 1 for an introduction). This area is characterised by a high degree of civic engagement in voluntary organisations, especially in sports and cultural associations. Some findings from other case study areas, mainly the Finnish and Scottish, will also be referred to.

The first aim of this chapter is to highlight gender differences in the types of voluntary engagement performed within selected sports and music associations. The second aim is to explore the gender aspects of networks related to these associations, and the power of the networks to influence the officially-promoted development course. Thirdly, the study focuses on the relation between individuals within an association and position in the labour market or in local business life, from a gender perspective. What are the advantages and what are the drawbacks? Finally, issues for potential voluntary work in the future are explored.

We take the point of departure at the level of the person because we find it meaningful to understand peoples' voluntary engagement and its consequences. It is then possible to relate this information to the level of the social in identifying certain practices and attributes related to the life of the association. Society is created through the socially-informed actions and practices of individuals. The experiences from some key informants can thus help us to understand society and to explain the relations – as well as the 'official talk'. Collectives are in turn also important for understanding and explaining the actions of individual people.

Methodology

The main empirical base for the study is semi-structured interviews with 15 women and 11 men, carried out in Leksand and Rättvik. In a few cases, results from a questionnaire will also be referred to (which, like the other questionnaires reported on in this book, was based on a purposive and non-random sample, in this case with 29 female and 24 male respondents). The people interviewed were selected in the expectation that they would be able to provide information on engagement, voluntary work and its dimensions within sports and music. We also looked for details from different levels within the organisations: for example in the Trotting Association in Rättvik five people were interviewed, including the managing director, the secretary, the marketing manager and voluntarily engaged women in the Ladies Trotting Club. We discussed 'non-engagement' too, in order to include people who had indicated in the questionnaire that they had no voluntary engagement. It turned out, though, that these people had been engaged in different activities in the past, and of course those who were engaged in present also had things to say about these issues. Finally, we wanted to hear general views about the role of voluntary work in local development.

The associations to which the voluntary engagement of the interviewees was related were, in the sphere of sports, Rättvik Trotting Association, Leksand Ice Hockey Association and Leksand Golf Club, and in the sphere of music, Leksand Fiddling Association, Leksand Folkdance Association, Dalhalla (an open-air opera stage outside Rättvik), *Bingsjöstämman* (an annual fiddlers' rally outside Rättvik), and Rättvik Dance Festival. Some complementary information about, for example, the number of members and board compositions was gathered directly from the associations.

The Gendered Aspects of Social Capital – a Theoretical Perspective

For Putnam (1993; 2000), social capital refers to features of social organisation such as building and sustaining networks, norms and trust that facilitate coordination for mutual benefit. Social capital thus develops as a (sometimes unintended) by-product out of interaction within groups. Concerning the end results of the building of social capital, Putnam focuses on the community level, claiming that there is a correlation on an aggregated level between lively participation in voluntary organisations and a flourishing economy. Dense, informal networks of civic engagement are seen as fostering norms of reciprocity and trust, which in turn facilitate cooperation, improve the efficiency of society and enhance the successful economic development of a community.

In contrast to Putnam's focus on social capital as a feature of social organisation and as a means to improve economic development of a community, Bourdieu's perspective on social capital is directed towards the level of the individual, not of course as a wholly independent actor but as a person habituated into certain modes of action. Social capital is thus the sum of resources that accrue to a person by virtue of possessing a network of relationships of mutual acquisition and recognition (Bourdieu and Waquant 1992, 119). Within any field, individuals struggle over resources and rewards, and their struggles are structured around their possession of economic, social, cultural and symbolic capital (Bourdieu 1977; 1991, in Shucksmith 2000). While Putnam underlines that social capital can only reside with groups, according to Bourdieu individuals can hold 'slices' of social capital, although these are dependent on their place and status in social groups rather than any intrinsic characteristics.

Bourdieu also relates the end results of the building of social capital to the individual rather than to the community level: the various forms of capital allow their possessors to gain power and influence. Putnam, on the one hand, conceptualises social capital as an unintended by-product of social activities that benefits also inhabitants who are not included in networks themselves but live in a community with dense networks. In Bourdieu's view, however, no accumulation of social capital is disinterested. However, it is important to point out that this interest related to individuals acquisition of social capital should not be interpreted in pure economic terms, such as in Adam Smith's sense of *self*-interest, which would reduce it to material benefit and the deliberate search for maximisation of monetary profit (Bourdieu and Wacquant 1992, 115–20). Instead, interest is historically arbitrary: each field presupposing and generating a specific form of interest, which may or may not have anything to do with the logic of the capitalist economy. In this chapter, we will consider both these perspectives on social capital – as a source of power on the personal or individual level and as a by-product of social interactions which is beneficial for the whole community.

There is a lack of research on social capital from a gender perspective. One of the few gender aspects highlighted by Putnam (2000) is that informal social connectedness is more common among women ('schmoozers'), while formal community involvement is more frequent among men ('machers'). But he argues that the entry of women into the paid labour force has shown that employment, not gender, is the primary key to formal community involvement (Putnam 2000, 94). Still, informal social connections are much more frequent among women than men, regardless of their job and marital status. 'Women are more avid social capitalists than men', claims Putnam (95).

However, Putnam does not actually go into the gendered aspects of social capital. Gender is a concept developed to contest the naturalisation of sexual difference – so just presenting statistical differences in sex distribution is not to have a gender perspective. Feminist theory around gender seeks to explain systems of sexual difference, whereby 'men' and 'women' are socially constructed and positioned in relations of hierarchy (Haraway 1997). In this chapter, we will not only consider sex differences in, for example, representation on management boards, but also the meanings and aspects of power related to gender roles.

Lowndes (2000) criticises Putnam and other social capital researchers for focusing disproportionally on male-dominated activities, like sports clubs, and excluding associations where women are in a majority, such as health and social services. Furthermore, she claims that not only are female-dominated formal kinds of voluntary engagement disregarded in the social capital research but that the social capital generated by women in the field of *in*formal sociability is also left out in the academic debate. Her main example is the area of mothers' informal child-care networks (including baby-sitting, the 'school run', the fetching and watching of children in club activities) – networks that are certainly characterised by mutuality and reciprocity.

Here, we would like to argue that not only have female-dominated formal voluntary associations as well as informal networks been neglected in the social capital literature. The gender differences in *roles* within associations, as well as in personal benefit for the position on the labour market derived from the voluntary engagement and in power to influence the course of local development have also been disregarded.

Exploring the Background to Voluntary Engagement

The Importance of Place for Voluntary Commitment

Generally in Sweden, people in small towns and especially rural areas are engaged in voluntary activities to a greater extent than those in large cities (Jeppsson Grassman and Svedberg 1999). In our interviews, participating in associations and other kinds of voluntary engagement was often put forward as a regional tradition. Rättvik and Leksand, but also Dalarna as a county, should be known as places where there is a strong will to engage in local development work and in the life of associations. It was suggested that this has to do with the historical peasant society, land ownership patterns and old traditions of village democracy.

> The old peasant society has, silently, continued to exist and brought people up ... yes, it lives on, its spirit of helping each other. (Martin, chairman of Leksand Fiddling Association)

The municipal government commissioner of Rättvik relates the question of voluntary engagement to land ownership, claiming people's commitment is greater in areas where the peasants traditionally owned their land, than in an area (in the northern part of the municipality) where the land was bought from the peasants by an iron works in the nineteenth century. Similarly, farmer-based networks in the Finnish case study area of Sotkamo, based on traditions of cooperation, are still important, even though the economic role of the primary sector has decreased. The sense of local identity, connected with the peasant relationship to place and the ownership of land-properties, gives continuity to farmer-based networks, as Lehto and Oksa noted in their research in Sotkamo.

Several of the interviewees in Leksand and Rättvik relate the high level of voluntary engagement to the old tradition of village associations, *byalag*. In almost every village today – and in Leksand we can find 80 villages – there is a village association. Traditionally, the villages were independent, and the gathering of *byalag* functioned as a kind of direct democracy. Village issues were discussed, decisions made, and maintenance and building work was carried out on a voluntary, community basis. We have not been studying these organisations closely but earlier research shows that today it is possible to find a scale of modern activities (between and within the associations) ranging from arranging parties, to trying to influence changes in the village concerning, for example, the local shop or day care, and the municipal strategies concerning local development in general (see Berglund 1998; Herlitz 1998).

> It is not unusual, even if I don't think it occurs to a great extent, that children join and help bake together, for example. It is rather the adults, but the children know

that the parents are there and they get 'reared' into the idea that you for example
go to the village house and help bake. (Martin, chairman of Leksand Fiddling
Association)

Today, most of the voluntary engagement in Leksand-Rättvik is directed
towards sports and cultural associations, rather than towards the village
associations.

There is a parallel between the tradition of village associations in Leksand-
Rättvik, rooted in a peasant society, and the crofting tradition in the Scottish
case study area of Skye and Lochalsh. Árnason and Lee (2003) argue that
crofting has provided small communities there with networking tools. In that
context cooperative labour, a strong meeting and committee structure and
deeply embedded social relationships provide the means by which crofting
takes place. The economic importance of crofting in Scotland, as well as of
agriculture and forestry in Leksand-Rättvik in Sweden and Sotkamo in Finland
has decreased, and yet are still of importance as *symbols* or representations of
a valued style of communal or cooperative working. These symbols can then
help mobilise or illustrate how rural development might work in the future
(Árnason and Lee 2003).

Another basis for the voluntary engagement in Leksand and Rättvik relates
to the size of place in general, rather than to specific assets in this area. In a
'small' place, more openness between people is claimed by our informants than
in large cities, where it is more anonymous. This means that people are more
inclined to help each other without money having to be involved. People are
also more likely to be asked to join an association in a small place, they said.
This could also mean that the moral pressure to join is greater than in a city.
One interviewee stated that the municipal government commissioner does not
write, 'now this piece of land has to be cleared for to save the village's view of
the lake.' Instead he just takes his saw and starts to cut the bushes himself, and
so encourages others to join in voluntarily. In this case it is the action (shifting
easily from individual to collective) rather than the discourse of engagement
that is most powerful, much as Cecchi found in his comparisons (Chapter 3).

Individual Motives for Voluntary Engagement

The desire to engage in voluntary work has been theorised by Jeppson Grassman
(1997), who argues that one way to interpret engagement is through the capacity
of voluntary organisations to attract people who have a will or feel obliged
to shape their own lives. This suggests we need to combine an analysis of the
collective and the personal aspects of volunteering.

If regional aspects are a part of the local discourse, it is still possible to find
different motives for engagement at the individual level. We have tried to find
out how people express themselves around their own engagement and voluntary
work and it turns out that both collective and individual aspects are important.

It could be described as a need for transforming the individual interest into a collective context. The engagement in the Ladies Trotting Club is one example. The personal interest in horses can be the basis for the engagement, but the possibility of feeling a sense of community and meeting people with the same interest as well as the desire to actually help out with the sport of trotting are also important ingredients.

While engagement in the Ladies Trotting Club is performed on a regular basis, the voluntary work at Dalhalla, an open-air opera stage, is an example of a non-continuous engagement. It is characterised by specific tasks limited in time, like taking the role of a parking guard, a waitress or a concert hostess for one evening. There is also a dimension of the expressing one's identity in taking on such a role. It could be interpreted as a combination of the individual and the collective, through the form of a social person. For our informants it is quite possible to support an event like a Dalhalla concert and a collective organisation like the village association[2] – but at the same time have the possibility of enjoying one's own interest (listening to music) and building up an identity through your specific role in the working team.

Several interviewees put forward the positive personal aspects for them that are related to voluntary work: one makes new friends, and it brings pleasure, a sense of community and satisfaction. Some suggest that people would feel good if they were more engaged, not least because they then would feel more 'needed' by the community. Again we see the importance of more-than-individual aspects of this engagement.

Gender Differences in Roles within and between Voluntary Associations

In this section we will explore, from a gender perspective, the aspects of hierarchy and difference within and between our case study associations. Even though the proportion of women in Sweden carrying out some kind of voluntary work (50 per cent in 1998) hardly differs from the share of men (53 per cent) (Jeppsson Grassman and Svedberg 1999), there are large gender differences concerning both the roles within the associations and the type of associations engaged in. Generally in Sweden, within voluntary associations more men than women hold board positions and carry out leadership and education tasks. Women more frequently work with information, opinion shaping, gathering money and direct help (Stark and Hamrén 2000). In Britain, similarly, men seem to be more likely than women to occupy committee posts (Lowndes 2000).

In terms of active participation, sports associations dominate the voluntary sector in Sweden. Men are more engaged than women, who in turn are over represented in social associations of various kinds (Stark and Hamrén 2000).

---

2   Members from the surrounding village associations help voluntarily at the Dalhalla concerts.

This gender difference exists in other countries as well; for example in Britain, twice as many men as women undertake voluntary work related to sports and recreation, while women are more active in the fields of health, education and social services (Gaskin and Smith 1995, in Lowndes 2000). Yet this does not mean that the female voluntary engagement in social services is as large in Sweden as in other countries. On the contrary, the main difference between the voluntary sector in Scandinavia and in other Western countries is that in Scandinavia, it is relatively weak in the areas of social services, health care and education of children, and strong in sports, leisure, culture, adult education and the labour market (mainly meaning trade unions) (Jeppsson Grassman and Svedberg 1996; Lundström and Wijkström 1997; Rothstein 2000).

Voluntary engagement in Leksand-Rättvik is directed mainly towards sports and cultural associations. Referring to the results from our interview study, we will begin by discussing the gender roles within some sports associations.

Voluntary Sports Associations in Rättvik and Leksand – 'Management by Sex Segregation or …?'

Erik, the chairman of the board of Leksand Ice Hockey Association (IAL) says that around 1000 people volunteer for the Association. Many are at the matches of the elite team (working in the kiosks and the car park for example) and others are around the boys' teams (taking care of equipment, helping at matches, selling lottery tickets etc). He estimates the share of women among these thousand volunteers to be around 40 per cent.

> If you look at stewards, there are a lot of women; they have their children playing in the teams and they make coffee, they stand in the kiosks. (Erik, chairman of the board of IAL)

Asked about whether those working voluntarily get some reward for it, he seems to disregard the large share of female stewards he has just mentioned, saying that 'Yes, we have a little party for all stewards once a year, when they can bring their wives'.

For a few years now, there has also been a female junior ice hockey team as part of IAL, with female leaders involved. However, neither this nor the share of women engaged voluntarily is reflected in the sex distribution in the boards of IAL. The main board consists only of men. Asked about whether there are women on the board, the chairman answers that now that there is a female team, there is likely to be a woman on the board in the future. He adds that ice hockey is a male sport, and that he has been working on getting a good mixture in the board, in which he has succeeded:

> You know I have been working a bit on this. There is someone on the board from the tourism sector, some from large firms; there are entrepreneurs, banks, insurance

Figure 4.1   A training session at Leksand ice hockey club

companies, the information sector ... so we have a very good mix. A very good board. (Erik, chairman of the board of IAL)

To the chairman, a good 'mixture' does not initially at least have to do with both sexes or other social groups being represented, only with a variety of businesses.

The tasks which the chairman associates with women working voluntarily in the IAL are dominated by 'traditional' or rather simple ones, such as coffee-making. However, as Stark and Hamrén (2000) point out, women in the voluntary sector often have positions with less power and yet greater responsibility than the men. Lotta, who organises the junior ice hockey team cups (the prizes), is an example of this. She says that the responsibility is significant, and that there is a lot of work which no one sees: 'They really don't understand how much time I have to spend on this.' She is a member of a board further down the hierarchical structure of the IAL, the 'youngsters committee'. There are in total two women and eleven men on this committee. Lotta says that it seems that because she is a woman, mothers and fathers of the boys playing come to her with their complaints rather than to any man, despite the fact that the issues might be better be directed towards a higher level in the association.

The main board of Leksand Golf Association also consists mainly of men: nine out of 11. Christina works voluntarily, mainly at the competitions. She also sits on a board further down in the hierarchy, the 'competition committee', which has two women and six men in it. She says that female representation in the boards is very important, as women have different demands to men (not least for physiological reasons) on for example length of the golf course. If there are no women in the boards, there is a risk that these demands be disregarded.

> Often it can be rather so-so with the distribution (of men and women). The men think, well the 'hags' should not have anything to say. But it is necessary for the sake of the women, well for the whole competition, that both parts are represented. (Christina, voluntarily engaged in Leksand Golf Association)

The Trotting Association in Rättvik is a non-profit organisation with the character of a company with employees but it also contains some other 'sub-associations' and voluntary work. The Ladies Trotting Club was founded about twenty five years ago. The fact that the Ladies Trotting Club is a pronounced sex segregated organisation should not necessarily be seen as a reaction to the otherwise very male dominated sport of trotting. The setting up of the club was more a result of some women's interest in horses and their desire to do something for the sport. It is thus an example of a more altruistic way of engaging for the collective good, but it is also an outcome of the personal interest in horses. They knew about another women's club at a neighbouring trotting arena and thought it was possible to set up a similar association. The aim of the club, as it was stated in the beginning, is to increase the cheerful atmosphere at the trotting track and to give something to the sport.

> There were so many things we thought we could do and we were very optimistic; that we could stand up for the trotting. And they still think it is very good that they have us to get back to if they want some help. (Evy, Ladies Trotting Club)

This aim can be seen as a traditional feminine aim – to take care or support others in the public or collective realm – but there are also other significant aspects in being a women's club. When we asked if it would be possible to let men into the club, we were told that would not be considered. If meetings and arrangements were mixed, it would end up with 'us serving them':

> If there would be chaps in the club then we would fall back to this 'we women will fix things'. Chaps wouldn't make the coffee or take the bread out or clear the table. Now we share those things. (Evy, Ladies Trotting Club)

Regarding male-dominated boards of sports associations, it is likely that existing networks are reproducing existing structures which means that a board with a majority of men will probably think in certain terms, and in which the women can also take part in a discourse that favours some masculine attributes.

Here Monika talks about selecting people for the main board of the Trotting Club:

> It could be a capable and pushing person. Doesn't have to be a self-employed, but it should be someone who has done something where you think that, well he is a good guy. You look a bit at what they have done before and that they maybe are used to sitting on a board. (Monika, Ladies Trotting Club)

One of the tasks of the Ladies Trotting Club is to search for both regular sponsors and sponsorship for their dedicated day at Rättvik Trotting. Such sponsorship provides an income for the trotting association, and the firms that advertise at the trotting track are mentioned in the prize ceremony. When the role of the Ladies Trotting Club is discussed with the manager (male), the marketing manager (male) and the head secretary (female), this aspect of the work of the Ladies Club is not mentioned at all. What is stressed is the character of the organisation as women who 'always are willing to help out' and are doing a good job.

> The Ladies Club is a very good association to cooperate with, very creative and often willing to help out, I would dare to claim always. There are other associations related to our activities which live in a much more anonymous climate and definitely dedicate themselves to wrong things. (Rolf, manager of the Trotting Association)

Even though women are directly involved in economic activities – such as finding sponsors for the trotting association – they are not considered to be part of the higher levels of the organisation. As such they are not represented in the board of the Trotting Association, and nor do they or the administrative people of the organisation connect them to economic contributions.

Differences in Sex Distribution in Boards between Large and Small Associations

What about female representation in the boards of the studied cultural associations? In the Leksand Fiddling Association, the board consisted of five men and two women. There has been a recent period of female chairmanship of the board. In the Leksand Folk Dance Association, there is a majority of women in the board: five out of seven.

It should be kept in mind that the difference between the total membership of these case study sports and cultural associations is large. While IAL has around 3,200 members and Leksand Golf Association 1,300, the Leksand Fiddling Association has only around 100 members and the Leksand Folkdance Association 80. In large associations the board often consists of people not necessarily directly involved in daily activities of the organisation. Rather, they have these positions as representatives of a certain business sector, for example, as was noted earlier. In these associations, it is usually at a lower level in the hierarchy that most of the voluntary work is done. This level involves rather traditional gender roles concerning the practical work, but could be seen as representing both men and women since both men and women are active here. The higher board level includes mostly men.

In smaller associations, though, it is usually the members of the board who carry out most of the voluntary work. Martin, the current president of the Fiddling Association, and Sofia, the former chairman of Leksand Fiddling Association bear witness to this. Sofia states that when she asked the other fiddlers at a weekly training occasion whether the association should accept an offer to come and play at some event, everyone was usually positive. But then it was her tiring task to call everyone and try to convince them to actually come, because at that stage people were often not that interested any more.

Figure 4.2   The Leksand Fiddling Association

Among the female respondents to our questionnaire, those who were volunteer members of a board turned out to be active mainly in non-commercial, small cultural and sports associations. Among the male respondents, membership in boards of large associations was more common, including not only sports associations but also semi-commercial cultural events like annual dance and music festivals. The men were also engaged in several boards, while the women mostly only in one. Generally in Sweden, women hold structurally lower positions in associations than men, and in cases where they are members of committee boards, this is more typically in smaller associations (Stark and Hamrén 2000).

This would mean that in cases where women have formal power in an association, it usually does not entail power to delegate and does not mean that the voluntary effort is restricted to the board meetings. Instead, in associations where women are typically board members, the position usually entails responsibility for many practical details in the association's daily activities and large personal contribution in terms of voluntary work.

Although often idealised, the voluntary sector is thus not free from uneven power relations, here manifested in gender differences in roles and influence within associations. As Stark and Hamrén (2000) put forward, it is difficult to influence these uneven relations through some kind of regulations from outside, as that would mean an element of *compulsion* being put on the *voluntary* sector.

What implications do these gender differences in roles within the associations have for the range of networks that both men and women are part of? Are there gender differences as to membership in networks which are 'cross-sectional', stretching from the voluntary sector to other societal sectors like business and the municipal authority?

'Pluri-positionality' and Cross-sectional Networks – Predominantly Male Phenomena?

According to Johannisson (2002) and others, the phenomenon of 'pluriactivity' is part of the new paradigm of rural economic development. It means that in entrepreneurial careers, periods as independent owner-managers are intertwined with other activities such as employment in the public sector or volunteering. We would like to call this phenomenon 'pluri-positionality', meaning that different positions in business, public and voluntary sectors can be combined simultaneously, or follow after each other in time. Using the concept of pluri-positionality also enables us to examine how people can use their position in one particular network to enhance another network in which they play a part.

Pluri-positionality seems to be a predominantly male phenomenon. Rolf, the manager of Rättvik Trotting Association, is one example. He used to be responsible for municipal tourism. This helped him to get the job as head of the Trotting Association and has also meant many advantages because of his earlier contacts and knowledge. The development of the trotting activities has been influenced by his way of thinking in terms of tourism and development, and we can also find cooperation with other bodies affecting the work within trotting activities. His friendship with a key person within the Leksand Ice Hockey Association resulted in a combined ice hockey and trotting night, and connections with different tourism-based activities are easily made, for example to the Classic Car Week.

Another example of pluri-positionality is a man who is self-employed in building and tourism, but has combined this work with chairing an annual dance festival association and with employment at the municipality to develop plans for a new business location area in Leksand and for a conference hall. The chairman of the board of IAL also practices pluri-positionality. He was the managing director of a large Swedish firm, and still has many board positions in the business sector. He says that he finds good use for his experience from the business world in the sports realm. And both his own network and that of the managing director of IAL, a former elite ice hockey player, have expanded thanks to this merger:

> He (the present managing director, former player) has provided me and I have provided him, so to say. We have interchanged our networks, simply. I have got his sports networks and he has got mine from the business world.'(Erik, chairman of the board of IAL)

In the Finnish case study area of Sotkamo, pluri-positionality is also practised. The director of a large hotel was a former member of the board of the baseball enterprise SuperJymy Ltd. When the municipality and the local Sports College founded the Vuokatti Slopes enterprise, the municipal industrial secretary first worked in it as a part-time director. When the municipality gave up its share of the company and it became privatised, he became one of the owners (Lehto and Oksa 2003).

There are female examples of pluri-positionality in our case study, but it is striking that it is limited to the voluntary and public sector, and does not include the business sector. One woman is employed at the Local Business Association in Leksand, and she is also a politician and a member of the board of the golf association as well as being engaged in the local branch of the Rotary Club. However, contrary to her male colleague, she has no positions in the boards of firms. Similarly, a female business advisor employed by the municipality of Leksand is not engaged in any boards in the business sector. She says that the offers you get, they are not from banks or similar; those are the 'uncles' clubs'. She is engaged in some small voluntary associations, like the women's football club, which have no connection to her paid work.

Some of the interviewees point out that there is a quite small group of people who appear in several official contexts in Leksand and Rättvik. One interviewee says that these same people with different positions constitute a group of perhaps twenty, of which the majority is men.

Cross-sectional Networks Influence the Officially Promoted Development Course

Raising the question of the 'pluri-positionalities' of people up to the level of associations, we can see that more informal and formal networking occurs between large male-dominated sports associations, the municipal and business sector, than between smaller, cultural associations (with a stronger female element in the boards) and the other sectors. One kind of semi-formal cooperation between ice hockey, local business and the municipal authority in Leksand is periodic job recruitment fairs, which have no equivalent in the cultural associations.

Linda, a member of the board of Leksand Folk Dance Association states that the municipality does not take advantage enough of her association's marketing value. Dalarna and Leksand are famous for their folk dance traditions, and yet the municipality puts forward the ice hockey in marketing contexts instead, she claims. When culture is used for marketing, it is rather seen as an accompanying element. For example, a folk dance team has accompanied the ice hockey team and the local firms to the match in Stockholm, and danced the midsummer dances around an artificial Maypole raised on the ice-rink.

Similarly, more informal and formal networking seems to occur among different sports associations than between sports and cultural associations. Rolf, who is the manager of the Trotting Association, underlines the importance of personal contacts and networks in creating new businesses and new forms of cooperation. He mentions some important names (all male) that have been crucial for new directions in developing the activities connected to trotting.

> Enormously important. Enormously important today. You build very much on relations, who it is that ... who you should talk to when it concerns hockey, or 'Vasaloppet' (a historical ski competition), or 'Classic Car'. Classic Car has a person called Roland Pettersson, who is very much connected to Classic Car. I think that my name is strongly connected to the trotting in Rättvik; Jonas Bergqvist to hockey, Mats Bud to 'Vasaloppet', or Rolf Hammar. It is enormously important. (Rolf, manager of Rättvik Trotting Association)

The examples mentioned here are connected to ice hockey, skiing and classic cars. Golf is another sport that is closely connected to the trotting. When it comes to events within the music or cultural sector the contacts are not so frequent – the cultural attributes are used in the prize ceremonies (women in traditional folk costumes), but cultural associations are not seen as economic partners.

Connections to certain key persons are not mentioned by the women from the Ladies Trotting Club. As they talk about the sponsors, it is in the sense that finding a sponsor will give you points. Points are awarded through the club for all kinds of voluntary achievements, and the more points one gathers, the less one has to pay to join club excursions.

In Sotkamo, we find similar dense networks between sports (in this case mainly Finnish baseball and skiing) and the municipal and business sectors (see Chapter 1). Cultural, female-dominated associations do not have similar cross-sectional networking. The dense cooperation between the baseball company SuperJymy Enterprise Ltd and the municipality in tourism marketing issues, even resulted in the former taking over the task of marketing the whole area of Sotkamo. The connection between Sotkamo tourist enterprises and the baseball is intensive. All the major enterprises in Sotkamo belong to the baseball VIP club. So in addition to having taken over the image marketing of the municipality of Sotkamo, the baseball club networks various local and national enterprises together.

Networks that have a wide spread in linking business, municipal and voluntary sectors are likely to have a decisive influence on the officially promoted development course. In Sotkamo as well as in Leksand, male-dominated, cross-sectional dense networks can be identified. These networks embrace the dominant development discourse, according to which sports (mainly baseball and skiing in the first case and ice hockey in the second) play, and should continue to play, an important role in local development. Even municipal representatives ascribe the ice hockey a decisive role: their metaphor for Leksand is, as noted in Chapter 2, a millipede, where the head is the ice hockey, the municipality the backbone and the local firms constitute the thousand feet. An alternative, less powerful development discourse, according to which culture and services are important ingredients, is embraced mainly by women in both Sotkamo and Leksand-Rättvik. In Sotkamo, the long term investment in sport and tourism for local development has been criticised by the female-dominated cultural activities sector.

Although the coalition in Sotkamo between sports and tourism could, on one hand, be seen as a cross-group solidarity which is favourable for the building of social capital, there is also an element of polarised solidarity. The coalition seems to exclude non-elite, grassroots sport clubs and cultural and village activists, among which are many women. In Leksand, it could be asked if there is a risk that the trust built between the ice hockey, the business sector and the municipality becomes particularised. According to Uslaner (1999), only generalised trust should be seen as contributing to the building of social capital.

The uneven gender possibilities that influence the officially promoted development course in Leksand-Rättvik are reflected in answers to the question of whether everyone has equal chances to influence development. While 12 out of 25 female respondents deny this, only two of 22 male questionnaire respondents did so.

The Relation between Voluntary Engagement and Paid Work

In this section, we focus on different qualities of relations between a person's engagement in a voluntary association and their position in or movement through the labour market. One dimension could be the possibility of finding a new job with the help of the networks built through the voluntary engagement. Another might be in the connection between the content of voluntary work and the present position in the labour market or local business life. A third aspect of this relationship has to do with questions of time and energy available for the individual to engage in both paid and voluntary work.

Voluntary Engagement Increasing Paid Job Opportunities?

Some of our female and male interviewees who were engaged at lower levels in the associations, for example the women in the Ladies Trotting Club and the male coach of a junior ice hockey team, do not readily see connections between their voluntary engagement and opportunities at the labour market. Others, however, engaged either on lower or higher levels, say that their engagement might conceivably lead to job offers. Lotta, who is in charge of the junior ice hockey cups at IAL says that because people in IAL might see that 'this is a good person, she gets things done', maybe she could be offered a job (she is currently a student). She meets many different people in IAL who work in different contexts on the labour market. Anne, who runs a taxi business with her husband also suggests that it is possible that her engagement in cultural associations and in politics could be useful for her chances in the labour market.

> I think so. When we have had a lack of work, then I have been thinking, I wonder what possibilities I would have on the labour market if I, so to say, would start knocking on doors and say that I am available. Yes, I think so, I think you establish those contacts. One realises; one wonders sometimes how some people can sit on certain posts, which one ... It is a large safety net; no one falls out once you have come inside. (Anne, leader of the music festival Bingsjöstämman)

In some cases, voluntary engagement in an association might lead to employment in that same organisation. One woman who was previously self-employed and a member of the board of the local business association is now employed there. The woman working voluntarily in a golf association was at the time of the interview expecting to be employed part-time at its office.

Looking the other way around – how the kind of paid work or self-employment engaged in affects whether on is offered a position and which positions in voluntary associations one is offered – was also a point of interest to our participants. To pick people for a board can be a very strategic choice. Some individual attributes are more attractive than others, and will probably differ between different organisations. As we saw earlier, a mixture of people holding high positions in different business sectors is desired in the case of the IAL board. For the board of the trotting association, to have people from the building sector, for example, is valued because of the practical knowledge and contacts such people can contribute.

> Often it is maybe a self-employed or someone who has some skills, whom you pick for the board, because that is always good, ... it is horse owners ... but you have

someone from the building sector, that is always good when we're going to build something. (Monika, Ladies Trotting Club)

The Relations (or Otherwise) between Voluntary Work and Employment

Whether there is any direct connection between the kind of voluntary work engaged in and present position in the labour market or local business seems to be a question both of gender and the level within an association that one volunteers at. Neither men nor women engaged in the lower levels in the case study associations had a direct connection between their engagement and their employment. On the board level, though, it is common to find that a person has been offered their position precisely because of their labour market position. The manager of the trotting association and the chairman of the board of IAL are examples of this. On the other hand, a position on the board level can be advantageous for a person's employment situation or own business. Gunnar, the chairman of the board of Rättvik Dance Festival, which organises an annual dance festival, is an example of a person having a more direct connection between voluntary participation and paid work. The contacts and knowledge about tourism he gets through his engagement in Rättvik Dance Festival he uses in his private business, a consultancy firm within construction and tourism.

> But this kind of voluntary engagement of course also creates contact surfaces for me and business opportunities. (Gunnar, chairman of the board of Rättvik Dance Festival)

The networks he became involved in and the knowledge of tourism gained thus increases his opportunities to get consultancy work. He has formal education in construction, not in tourism, but:

> I have learnt very much about how tourism functions; without being educated, I know quite a lot ... All knowledge is perishable, and it is about having a circle of contacts, that is what I get, 'social competence'. (Gunnar, chairman of the board of Rättvik Dance Festival)

He says that the has 'steered' his daily networks: 'that is what you have to do to get things done.' Previously, he was member of a voluntary association which drew up plans for the construction of a Lake Siljan Hall, a big event and conference hall in Tällberg. At the time of the interview he was employed as consultant by the municipality, making more detailed plans for that same hall.

Among the respondents to the questionnaire, a direct relationship between the paid work situation and the kind of voluntary work engaged in was more common among the men than the women. Men's participation in the labour market and voluntary sector seems more woven together, to the extent that the engagement in a voluntary organisation and the employment (or self-employment) becomes hard to separate. One link comes through the sponsors of the ice hockey team, who are members of the so-called 'Gold Club' in IAL.

Is this phenomenon of fluid borders between paid and voluntary work only a question of level in the association and of how high the position in the labour market or business is? That is, do gender differences just reflect the fact that men are overrepresented in higher positions in the labour market as well as in associations? What about women holding high labour market or business positions? From our data, we can only state that one of the few women we studied who, as municipal business advisor, holds a high position on the labour market, is not engaged in boards in the voluntary nor business sector that are related to her paid job. She says that she has consciously tried not to engage in several boards related to the business sector, because if she gets very involved in them, she could no longer be a good, neutral business advisor.

Voluntary Engagement as an Arena for Restoration from Paid Work

But in other cases we can also identify a quite different quality of relationship between voluntary activity and a paid job. It consists of restoration from stress: the participation in voluntary work enables individuals to 'shut off' their paid work, helping to restore themselves and get new energy to meet the demands of the labour market. Sofia, former chairwoman of Leksand Fiddling Association and employed full-time as social welfare secretary, says:

> No, my engagement in the fiddling association has nothing to do with my job. It is very different; it is another kind of communication. There you communicate through the music, through playing with others. At work, there you talk and solve problems and have a lot of writing work and a lot of social relations and so on. The music gives me strength ... Though awake, you get an incredible rest. (Sofia, former chairwoman of the board of Leksand Fiddling Association)

Similarly, Therese, the business advisor, says that if someone like her is engaged in an EU project, then 'you are either burnt out or at least burnt'. You have to have something else in life that makes you shut off from work so that you can calm down:

> My engagement in the ladies football team, that is because I have to get another life, not just live in this (work) life. You have to interrupt the work with something totally different and to meet people who do not have similar thoughts as I have. Or else you 'get blinkers'. (Therese, municipal business advisor)

From the perspective of restoration, it is rather an advantage if there is no connection at all between voluntary engagement and the paid work. The possibility that voluntary work could allow for network building, which can be advantageous for labour market position in terms of career or better business opportunities, becomes irrelevant. In the interviews, the idea of restoration was put forward mainly by women. As women in Sweden often have a larger total workload than men, being engaged in paid work and yet having the main responsibility for the domestic work, it might be expected that this aspect *is* even more important for women than men.

Time Restrictions on the Combination of Voluntary and Paid Work

One further aspect of the relationship between voluntary and paid work is the difficulty of combining them due to time restrictions. Little (1997, 201) found that while women living in the British countryside are getting more involved in paid employment, it is still assumed that they will participate in voluntary work, which can be hard in terms of time and geographic positions. In some cases, women's commitment to voluntary work left them no time to take a paid job. In our interviews, we found no examples of the latter. However, there are women who have reduced their voluntary work because it becomes too hard to combine a large voluntary commitment with demanding paid work. Sofia, from Leksand Fiddling Association, says:

> It (the involvement in the fiddling association) has given so much; it has been great fun. Many great years. But now towards the end, it got to be more effort than it gave me back. When it empties your own resources, you give and give a lot at your paid work too … Oh, many stones were unburdened when I quit this chairwomanship. It was a too heavy burden of responsibility lying on my shoulders. I have enough responsibility at my paid work. (Sofia, former chairwoman of the board of Leksand Fiddling Association)

Similarly, Therese, the business advisor, says that she had to quit the chairwomanship in a horse riding club because there was insufficient time to commit to it as she took up paid work.

The Future of Voluntary Engagement

Putnam (2000) is concerned with what he argues is the diminishing level of voluntary engagement among Americans, leading to declining social capital. What are the important aspects for the degree and direction of future voluntary work that can be identified in our case study?

Several of the interviewees, both female and male and both in cultural and sports associations, underline that it has become increasingly difficult to get people voluntarily engaged, and they are worried about the future of activities that rely on voluntary work. Similar concerns were expressed by members of sports associations in the Finnish case study area. When one constantly has to 'nag' at people about helping out, it becomes a compulsion which is the contrary to the principle of volunteering. At the same time, it is emphasised by municipal representatives in Leksand and Rättvik that more voluntary work is needed in the future. The municipalities will not be able to provide all the services they do today, as they face an ageing population.

From Fixed to Fluent Engagement?

There is a lack of young people in the associations we have studied. In the fiddling and folkdance associations, people in their 50s dominate. One woman says that 'we have lost a generation' and the reason is that the people in their 50s and 60s have been satisfied within their own age-segregated networks. Their children have usually grown up, so there is more time left to engage voluntarily. Retired people seem not so common as expected. One interviewee says that his experience is that pensioners, though having a lot of time, never have time to join things – 'they have learnt to say no'.

One reason given for the difficulties of getting people engaged in traditional voluntary work, carried out on a regular basis, is that many kinds of *paid* work today are more and more project-orientated, demanding a lot of creative thinking and working long days. This leaves less energy for voluntary work.

> This is the problem, for your and my generation. We don't have the scope to work voluntarily in these associations. Today, you are expected to be engaged in your children's associations, in school, you should be active here and there. I see today in association life – look at the old heritage association, 55+ isn't enough, it is 65+ or 70+. Who takes care of the old heritage association in ten years? Who arranges these activities? Who arranges the 'culture walks' in Rättvik next year or in ten years? We don't have time to devote to these voluntary jobs. It has to be changed in society, but it is a big problem. A great deal is handled on a voluntary basis in the tourism sector, but those born in the 1960s and later do not have the time that those born in the 1940s and 1930s have. We should be aware of it, it will come. Will people call for the municipality then, or for the state? It will come, it's just that it is 10-15 years ahead. (Rolf, manager of Rättvik Trotting Association)

Rolf means that less time is available for younger generations to engage in traditional activities that relies on voluntary efforts. Besides more modern forms of voluntary work, like helping out in one's children's' organised leisure activities, engagement might have become more directed towards activities of personal or 'hobby' interest and towards the nearest surroundings.

Wijkström (1998a) uses the term 'new volunteers' to refer to the tendency among certain Swedish organisations to 'recruit volunteers for more limited assignments and projects'. The consequence is that the relationship between an organisation and its volunteers could be characterised as a contract rather than a traditional membership. Perhaps, he argues, such a relationship could be more suitable for the participation of the people of today: 'Without committing oneself to all the different activities or to the more administrative chores of the association, it is possible for a person to focus on the particular task or fields in which he or she is interested' (Wijkström 1998a, 65). This may explain why the engagement at concert nights in Dalhalla is so popular, and why it is said to be easy to get people work voluntarily at the elite ice hockey matches. The tasks are specific and it is easy to take the role of a parking attendant, a waitress or a concert steward for one evening.

Such a trend towards more flexible engagement, related to a personal interest, seems difficult to combine with the expectations from the side of municipal authorities for parts of public services, performed on a regular basis, to be taken over by voluntary workers in the future.

Traditional Gender Roles Preventing Rejuvenation of Voluntary Associations?

The desire within associations for continuity and a stable (or even increasing) membership is not disputable. But what about critical self-examination within the organisations? In the Ladies Trotting Club the interviewees explain the unwillingness to participate in terms of the hectic life of young women and problems with combining child care and full-time working. But another side of the coin is the character of the organisation. Are young women of today really interested in serving a male dominated organisation like the Trotting Association – which is the purpose of the Ladies Trotting Club – as a parallel task to their own personal well-being? This may be the case even though their activities within the club are highly appreciated:

> We know that they [the men within the Trotting Organisation and the Board of the organisation] think we are doing a good job.

> The Ladies Club is always willing to help out.

> Their work is appreciated.

But could it be that younger women are not satisfied with being seen as a service club, however important that might be? They are, perhaps, looking for engagement in more gender-equal forms.

The Ladies Trotting club has to be self-reflexive about its role. This is underlined both by Rolf, the 'outsider' and non-engaged manager of the Trotting Association, and by the women engaged in it. Rolf points out that the women in the Ladies Club get older and older, and that there are problems in recruiting young ones. But the role of the 'mothering' (or in this case 'fathering') organisation can also be taken under consideration. The question of female influence should not only be about getting the women engaged in the Ladies Club or to encourage attendance at the races through arranging 'ladies evenings', but could also be seen as important in the whole structure of the organisation.

One study in Skellefteå in Sweden showed large differences in what was expected from girls and boys in junior sports associations. The expectations for the girls' sports results were low, while the expectations on their voluntary contribution in the associations were high. The opposite was the case for the boys (Sveriges Ungdomsprojekt 2000, in Stark and Hamrén 2000). Traditional, uneven gender relations which are built into the structure of voluntary associations might thus be hindering the rejuvenation of memberships. Such relations consequently now challenge the conditions for these associations to function in the future.

Commercialisation – Threatening the Desire to Engage Voluntarily?

Both the trotting and ice hockey associations attempt the difficult balance between commercialisation and professionalisation on the one hand, and voluntary work on the other.

> We are a branch today. Many are employed … The balance between the necessary business elements and the voluntary work, that is the difficulty for us today and in the future. Before, it was greatly on a voluntary basis, but today there is so much money in the horse sports, so it is not possible to develop on a voluntary basis. And this is a difficulty; the Ladies Club as a voluntary organisation, strongly connected here; then you have the necessary business elements which we work with, companies, arrangements, sponsors, we take fees, entrance fees … what it costs to enter here in contrast to the voluntary efforts. That is the difficulty. We are in the midst of it. (Rolf, manager of Rättvik Trotting Association)

In Sotkamo, the increasing commercialisation and professionalisation of baseball has brought about tensions between the earlier voluntary work and the new orientation towards money-oriented entertainment. The tendency towards seeing the voluntary activities of sports clubs as part of the local economic activity has to some extent been criticised by women. The women there suggest that before sports became business, they participated eagerly in voluntary work.

There might thus be a danger that the commercialisation of sports, which can have *direct* positive effects on the local economy, has long-term negative effects on the building of social capital (commonly seen as having *indirect* positive economic effects).

The Importance of Place for Future Voluntary Engagement

Some of the interviewees who related the large voluntary engagement in Leksand-Rättvik to the old peasant society (whose spirit of mutual help is seen to still influence people's attitudes and behaviour) express fear that this spirit is declining. Similarly, in the Scottish case the decrease in crofting agricultural activity may be undermining the symbolic power of crofting as characterised by sharing and cooperation. This decline can in turn be seen as undermining the effectiveness of previously existing networks, like those of referrals amongst Bed and Breakfast proprietors (see Chapter 7).

However, some more positive views on the future of voluntary work are also expressed by our interviewees. As noted, it is put forward as a probability that in general terms fewer services will be provided by Swedish municipalities in the future. Facing such a scenario, rural areas like Rättvik and Leksand have an advantage precisely because they have a tradition of large voluntary commitment – it is their strength. It is also the case that small places, especially villages in rural areas, already get less public services. Thus, their inhabitants are more used to arranging things on their own.

Concluding Remarks

One critique of earlier research is that civic engagement as an indicator of social capital is commonly treated on an aggregated level that does not allow for differentiation according to content, roles and meaning. Putnam (2000) measures civic engagement with the help of, among other indicators, attendance of clubs and membership of associations. We would like to argue that when no distinction is made between the monthly attendance in a large sports club board and the weekly gathering of clothes and money for the Red Cross, for example, important differences in the quality of social capital are disregarded.

A platform for studying civic engagement in voluntary associations is the knowledge of the different roles which can be performed in connection to an association. We have found the following five forms of 'using' the associations and their activities. What makes the situation complicated is that one association can be used for all these practices but with differences among different people. It is thus not meaningful to try to characterize an organisation according to prevailing practices – it would be more accurate to talk in terms of different practices within the 'associational life' as a whole.

Associational Practices

1   *An arena where people perform a special activity (e.g. they play the fiddle on Tuesdays but do not work selling lotteries or as parking guards)*

This practice is often wrongly denoted as voluntary work. Both men and women perform this associational practice. The impression from the interviews is that this kind of practice, differently from the voluntary work, is not in danger of decreasing. There are seldom direct connections between this practice and the labour market position,[3] and the prevailing relation might be the restorative aspect, i.e. the performed activity allows for the 'shutting off' of the paid work for a short period of time.

2   *An arena where people work voluntarily; continuously or more 'ad hoc'*

In larger associations this practice is found mainly on the grassroots level, while in smaller associations it is common among the board members. Both women and men are found here. The practices are often characterised by traditional gender roles, especially in the case of lower levels in large associations. The networks which are built from this kind of practice are seldom beneficial for a person's labour market position or private business. On the contrary, there is usually no connection at all between the content of the voluntary work performed and the individual's paid work.

We would like to argue, though, that the kind of social capital derived from this practice is the one that most corresponds to the view of social capital as a feature of social organisation able to benefit the whole community (a view embraced by Putnam 2000; Coleman 1990; Warren 1999). Rather than mainly resulting in increased personal power and influence for those engaged, the networks of mutual trust and cooperation related to this practice might have 'spill-over' effects benefiting also inhabitants who are not engaged themselves.

Furthermore, it is easy to join in this kind of practice. In comparison to the board level of large, semi-commercial associations, there is less requirement to have a large stock of one's own social and cultural capital before being allowed to enter. These smaller 'entering restrictions' mean that there will be a greater mixture of people in terms of their background. Here we find people with high and low levels of education and labour market positions.[4] The trust that is built can therefore be assumed to be more generalised (embracing people who have different values and different backgrounds) than the trust built within the networks on the board level in large associations.

However, there is a trend towards more 'ad hoc' engagement, at the cost of regular commitment. For example, the post of the secretary in the fiddling association was recently vacant for some months; something that has not previously happened. This trend might change the prerequisites for the functioning of mainly small, non-commercial associations in the future.

---

3   We do not mean that there is no relation at all between the labour market position (reflecting class) and the kind of association activity chosen – rather that there is no connection between the actual practices performed in the association and at work, nor between the networks related to the association and to the work life.

4   This is not to say that all social classes are equally represented. We are aware that there exist class differences on the national level regarding membership in different types of associations. However, especially in team sports associations like ice hockey and football, differences in class representation are insignificant (Statistics Sweden 1993).

There is also the question of the future of voluntary work that is characterised by traditional gender roles. Generational shifts are needed within the organisations but a possible problem is that the traditional gender roles prevent rejuvenation of the member force.

3 *An arena where people have a position of responsibility (e.g. on a board) without being employed*

In larger associations, this practice is dominated by men. A direct connection to the labour market position is most likely to be found in relation to this practice. It can occur that a person got the position in the board precisely because of his or her labour market position. On the other hand, a position on the board can be advantageous in terms of career opportunities in the labour market, and the network built can be used in the one's own business. This practice is often interwoven with the person's labour market practice, making the voluntary engagement and paid work difficult to separate.

We would like to argue that the kind of social capital related to this practice is more often built on particularised trust than that built on grassroots engagement. The networks related to this practice are in many cases exclusive in terms of gender (commonly male-dominated), but also in terms of class and, we think, ethnic background. The personal, pre-existing cultural and social capital is often decisive for the chance to get into this kind of practice and its related networks. While the social capital built through this practice might also benefit the community as a whole, to a larger degree than the social capital related to the grassroots level it seems to correspond to the view of social capital as an individual's resource (Bourdieu and Wacquant 1992). Personal benefit and power can be gained through the networks built.

4 *As a place for employment, either a high or low position, such as the head of an organisation or a caretaker*

Many voluntary associations also have employed staff. There are examples in our interviews of people who combine this employment with voluntary work in the same association, while others have a voluntary commitment only in other organisations. People practising 'pluri-positionality' can be seen to change between periods employed at high levels in associations, for example as managing director or sales officer, and employment or self-employment in the private sector.

5 *As a platform for business contacts, e.g. people are 'visitors' and invite business clients to an ice hockey match*

Several of the interviewees speak of the ice hockey matches, the trotting competitions and the golf as places where people with whom one has business contacts can be brought. In the Finnish case study area, the Katinkulta Spa

Hotel and its sports activities, such as golf, have brought together the local elite and visitors to Sotkamo. It can thus be said that the associations' activities act as platforms for informal business contacts for the sustaining and strengthening of business-related networks. A precondition for these platforms to function is often voluntary work performed at the grass roots level.

In conclusion, it is crucial to understand the nuances of the various forms of associational practices and the kinds of social capital built in relation to them. Furthermore, it is important to consider *which* networks have the power to influence the officially promoted line of local development. It cannot be taken for granted, as it seems to be by Putnam, for example, that the existence of a large stock of social capital will always be beneficial for the whole community. We would concur with Shucksmith (2000) in his critiques of the treatment of social capital as a collective good in theories of endogenous development. This 'masks the way in which these assets are appropriated by those who individually already have social and cultural capital' (Shucksmith 2000: 6).

According to Bourdieu (1977; 1991 cited in Shucksmith 2000), within each field the dominant class, through 'symbolic violence', defines what social relations are valuable and what knowledge is legitimate. In our case study context, this conception might be translated into a male-dominated group holding positions in cross-sectional networks, which defines what constitutes the community interest.

The cross-sectional networks in Leksand-Rättvik, linking the voluntary, business and public sector, *do* play an important role and are certainly one of the keys to the understanding of the relatively successful economic development of the area. And yet, the existence of dense, cross-sectional networks *per se* should not be seen as an automatic key to successful development in a broad sense, i.e. including non-economic aspects. On the contrary, there is a danger that such cross-sectional networks are exclusive in terms of gender and class, for example, and that they have a disproportionally large influence on the officially promoted development course, suppressing alternative development discourses. We feel that the crucial question for rural development of how to build social capital in the collective sense of Putnam, and yet without excluding or doing symbolic violence towards marginalised groups in Bourdieu's sense, remains.

Chapter 5

Identity-building in Regional Initiatives for Rural Development: Comparing Ireland's Lake District and Norway's Mountain Region

Torill Meistad, Frances Hannon and Chris Curtin[1]

Introduction

In Europe there is an increasing interest in regional cooperation for industry and community development. The concept of 'integrated rural development' emphasises the territorial dimension and cross-sector application in the reform of European rural policy. This implies an upgrade of rural development as the second pillar of the Common Agricultural Policy with a purpose to diversify the rural economy, and improve the quality of life. Among other issues the Cork declaration (1996) mentions 'promotion of culture, tourism and recreation' as an element of rural development policy. Of special interest for this paper is the exploitation of local development potential and the involvement of local people which indicates a focus on endogenous rural development processes (Van der Ploeg and Long 1994; Ray 1999; Bruckmeier 2001).

In this chapter we will present and compare two European regional initiatives working for rural development. One is the Lake District Enterprise, which is a community enterprise in western Ireland. The second is the Mountain Region Council, consisting of inland municipalities in southern Norway. Both regions are experiencing the threat of socio-economic decline, but are seen to be succeeding in maintaining population and employment compared to similar and neighbouring communities. They are also among the many remote rural areas in Europe experiencing the challenge of the global economy, encouraging their

1   The authors thank the people in the Lake District and the Mountain Region who participated in this project. Hannon and Curtis thank in particular the members and managers of Lake District Enterprise who gave generously of their time and unearthed relevant information on more than one occasion. Louise Kinlen and Pauline Joyce contributed to the Irish research, while Reidar Almås, Svein Frisvoll and Johan Fredrik Rye also worked in Norway. The authors are grateful to members of the RESTRIM teams in other countries who have shared ideas, experiences and information.

communities to diversify industries and maintain welfare in order to attract and provide income to future inhabitants. The two cases were selected because they have been relatively successful regions regarding social and economic changes during the last couple of decades. Both areas have experience with regional initiatives for rural development which we find fruitful to compare.

In our search for explanations for this relative success, we are focusing on regional identity as social capital in rural development. Social capital is often referred to as 'the glue' that binds people together. It has been explained as a feature of social process, including the building and sustaining of networks, norms and trust that facilitate coordination and cooperation between groups for mutual benefit (Putnam 1993). We intend to explore regional identity as a type of social capital. Regional identity is of special interest when establishing a new territorial unit for the purpose of a joint community development initiative. Our study will focus on the process of building regional identity as a part of the initiatives in Ireland and Norway. Our research question is as follows: how is regional identity-building involved in rural development initiatives?

By taking a comparative approach we can explore how initiatives with a similar purpose vary in contextual conditions, strategies and results. They illuminate the complexity of regional identity-building. Within the two case study regions we expect to find similar challenges for community development, and also similarities in territorial identities. Due to differences in national governance, however, we expect to find variations in organising and financing the initiatives. Of special interest is whether these variations imply differences in the strategy of regional identity-building.

Social Capital and Collective Identity

Robert Putnam underlines social capital as a collective resource. Similarly, we take regional identities as phenomena that exist only by collectively sharing and repeating histories, symbols and narratives. Describing regional identity involves identifying associated symbols and social practice, as well as considering the core strength and consciousness of the identity in question (or what might be seen as the essential qualities) (Paasi 1986; 1996; Lysgård 2001). Paasi writes that a region in this context includes 'the production and reproduction of regional consciousness in the inhabitants and those living outside the territory, and also material and symbolic features of the region as part of the ongoing process of social reproduction' (Paasi 1996, 99). From this broad definition, therefore, in this chapter we will trace the symbols and content of regional identities, with particular reference to the changes in social practices of regional identity-building.

The analysis is supported by recent trends in regional geography, defining regions as socially and culturally constructed categories, dependent on history and formed by discourses. This underlines the dynamics of social geographical units, which we find useful in our analyses of regional identities as part of rural development strategies. Such a geography also engages with questions concerning the formation of regions and regional identities. Gilbert (1988) identifies three main approaches: the region as a response to capitalistic processes, the region as a focus of identity, and the region as a medium for social interaction. In our study the relation between the first two is of special relevance in exploring the dynamics between external challenges and internal reactions in rural community development.

The new regional geography tradition highlights the dynamics of regions and considers regions (and regional identities) as part of an ongoing historical process. The 'life cycle' of regions as socio-spatial communities is theorised by Paasi (1996) as the institutionalisation of regions. Regions are seen as spatial units that have been produced socially and culturally to become part of the territorial system. They exist for some time in social and cultural practise and discourses and then disappear in a continual process of regional transformation. (Taylor 1991; Paasi 1986; 1996). Ongoing community processes will always contest and change regional units and identities. This perspective is relevant in the study of the Mountain Region Council and the Lake District in the dynamics of creating regional identities.

We also find support in studies of local identity as part of bottom-up initiatives in local communities. The bottom-up perspective helps to explain a common regional identity in scarcely populated areas as a counter-action against negative statistics on population and economy (Wollan 1994). A regional identity in such areas can also serve to highlight various advantages often associated with 'remoteness', such as active non-profit organisations and high-quality environment (Veggeland 1993; Paasi 1990; Chapter 7 of this volume). Some actively promote such qualities in slogans, official emblems, festivals and so on (Meistad 2001). This research engages in the study of subjective qualities and might be seen as supplementary to the more materialistic community planning tradition.

Paasi's investigations of four Finnish regions brought to light the construction of territories as part of continuous change or a transformation taking place in the spatial system (Paasi 1986). Among other things, he problematised the links between a group of people and a bounded region by questioning basis of the regional consciousness (Paasi 2001). Among those following this challenge is Hans Kjetil Lysgård, who found no institutionalised spatial practices or regional identity after a long-term political cooperation across national borders in the Mid-Nordic region (Lysgård 2001). A further useful study in this vein explores the dynamics of identity construction and deconstruction of the 'Green Heart' region in The Netherlands (Van 't Klooster et al. 2002). This example illustrates how a metaphor became articulated as the environment changed, and later became internalised as part of regional planning policy.

Materials and Methods

Researchers in the respective countries collected information about the case study regions, further introductions to which are in Chapter 1. The analyses are based upon three sources of information.

Firstly written information, covering statistics on the demographic, economic, political and social development situation within the regions and in the national and regional political economy. This information highlights challenges and places the two case studies in context.

Secondly a questionnaire was carried out in the two case regions, directed towards people involved in development activities. This covered the issues of important factors of community life in the area, development activities and organisations in the region, and the respondents' role in development networks and associations. The questionnaires were informative for characterising the qualities associated with regional identity.

Thirdly, documentation and semi-structured interviews about the regional development initiatives: the background to them, actors, activities, conflicts and experiences. This material helps to provide understanding of both the strategies and the results of these initiatives.

The case study strategy was chosen to combine the explorative, descriptive and explanatory qualities relevant for the research questions (Yin 1994). The multiple data sources involving surveys, interviews and document analyses covers the requirement of both in-depth and overview information, highlighting variations and changes over time. The two cases chosen are considered typical for European communities actively working for rural development by creating regional units. The limited number of case studies gives us the chance to intensively analyse the relationship between identity and community change.

In this chapter we will present the two case study initiatives and the territorial regions in which they operate, compare their regional identities, and compare the regional identity-building activities. Before concluding there will be some comments on themes to consider in analysing the building of regional identity.

The Mountain Region Council and Norway's Mountain Region

The Mountain Region Council is a political and administrative body established in 1995. It is the driving force for expanding regional cooperation in a highland area in the inner part of southern Norway, close to the Swedish border. The Norwegian Mountain Region includes nine municipalities, covering parts of two neighbouring administrative counties. The region is among the most sparsely populated in Norway with less than two inhabitants per km² (totalling about 25,000 inhabitants and an area of 14,200 km²). Most of the area is above tree growing altitude, and only a small percentage of the land is suitable for cultivation and housing. There are two regional centres, the towns of Tynset and Røros.

Figure 5.1   The Mountain Region, Norway

Farming, forestry and hunting are still important sources of employment in the region. The mining of copper had a 300-year history until the last mine closed in 1986, and the mining town of Røros is now a UNESCO World Heritage Site. These traditional economic activities are still at the heart of the regional identity. The Mountain Region has a history of struggling against a harsh inland climate and poor living conditions. Summer dairies in the high pastures still operate, making the most of marginal resources. On the other hand, the church of Røros is a symbol of a period of wealth that came from the international trade of copper. Even if the living conditions of today match national levels, the culture and mentality is still embedded in 'a collective oriented social radicalism mixed with strong Protestant hardworking ethics and cultural conservatism' (Almås 2003, 176).

Inhabitants of the region who took part in our questionnaire appreciate the quality-of-life aspects of their surroundings. They value the nature and scenery, and clean air and opportunities for outdoor life. They also highlight the good environment for raising children and the social life of the small communities. In addition, a good job market, possibilities for starting up businesses, good public services and low living expenses are among the positive qualities mentioned by inhabitants.

In Norway there is an active local democracy, and a municipal council is elected every fourth year. The municipalities are responsible for children at pre-school age, primary schools, services for elderly people, territorial planning, sanitation and so on, and are responsible for many local jobs. In addition, more responsibility for industrial development has been gradually decentralised to county and municipal level. Public administration and services employ one-third of the labour force.

Incentives for new regional cooperation in the early 1990s included the challenges of growing globalisation in the economy and a weakening of national policy for the peripheries. Cuts in agricultural subsidies hit marginal farming, manufacturing industries were affected by low cost production overseas, and national policy changed to reduce growth in public costs and employment which expanded heavily especially in rural and peripheral areas of Norway during the 1980s. Even within their respective counties, the municipalities in the Mountain Region are peripheral. Politicians of the neighbouring municipalities recognised the need to work together for the interests of small and remote communities. Typically the Mountain Region Council lobbies for better roads, the up-keep of train and airport services and reduced taxation for enterprises in peripheral areas, with a number of successful results.

Purpose of the Mountain Region Council:

The Council shall:
– contribute to the Mountain Region remaining a good place to live;
– improve the accessibility, market orientation and base of knowledge in the region, to increase the variety of employment possibilities and versatility in economic life;
– create sustainable development;
– represent the involvement of the municipalities at the regional level.

The Council has since discussed possibilities for shared public services to save money, but this has been met by internal resistance among some members. Initiatives for regional industrial networks, such as in forestry and tourism, have not had much interest so far. On the other hand, decentralised university courses have been a success, with courses ranging from business and administration to health and education. The new strategy is inter-municipal planning for industrial development, for new housing and for cultural services. The attitude that every municipality should have all services within its own borders is being replaced by a willingness to supplement each other, with the aim of creating better services for the inhabitants of the whole region.

The Council grew out of a four-year period of irregular meetings of local politicians and head administrators, representing each of the municipalities and counties. The establishment of a secretariat to administer common initiatives indicated the beginning of the formal regional network. The purpose is to coordinate development politics in the member municipalities. The Council has no executive power of its own, and the discussions are only advisory regarding the municipal boards. The secretariat is financed by membership fees from the municipalities (a small share of the local tax revenue) and support from the county administrations. Projects are financed individually.

There were eight municipalities involved from the start in 1995. Two years later a neighbouring municipality in the south also joined. In 2002 two of the most northern municipalities decided to resign from the council. This was the consequence of a two-year process of self evaluation and a readjustment to the strategy of the regional council. The Mountain Region Council is also part of an alliance with four neighbouring inland regional councils which coordinates efforts within the national political network and has been particularly active over the issue of reduced employer taxation in inland areas.

Regional councils such as that of the Mountain Region are a new type of initiative in Norway. They are part of a well-established system of democracy, public service administration and supervision, which exists at three levels: municipality, county and national. Over the last three decades in Norway more responsibility for public services has gradually been devolved to municipal level. This is also the case for industrial development. Even if there is a national policy and institutions for supervision and funding at national and county level, initiatives are coordinated and related to community planning at the level of the municipality. The new regional councils have gradually emerged from previous experiences of cooperation among politicians and municipality administrations. The regional councils are thus partly a response to the requirements for specialist competence to deal with the need for coordinated planning in an increasingly mobile population.

The Lake District Enterprise and Ireland's Lake District

Ireland's Lake District is located in the north west of Ireland. It covers parts of south Mayo County and a small part of north Galway. The area includes 26 local communities, with Ballinrobe as the regional centre. Agriculture and small industries still dominate. Compared to the Norwegian case study region, the Lake District has about half the population (approximately 12,000 inhabitants) and less than one fifth of the area (850 km^2). Still, the region is among the most sparsely populated areas in Ireland (16 inhabitants per km^2).

The lakes that characterise the region cover nearly half the area (400 km^2). Lough Mask in the north and Lough Carrib in the south are the largest. The natural qualities are highly valued both for recreation and as an attraction

Figure 5.2   Post office in Ballinrobe, in the Irish Lake District

for tourism. The possibilities for angling, in particular, are mentioned in questionnaire responses. The Lake District is a place of relatively undiscovered natural beauty, and is described as 'one of Ireland's best kept secrets'. Quality of life in the Lake District also includes the rural, traditional way of life in the dispersed villages. As a part of western Ireland, it is considered the most authentic and oldest of Irish regions (Whelan 1993).

A long-lasting decline in population was supplemented by an additional wave of emigration to urban areas in the 1980s, during the period of economic growth when Ireland was popularly known as the 'Celtic Tiger'. As a consequence the number of primary school enrolments dropped dramatically, for example. Since 1996 the population has increased again, partly due to in-migration and counter-urbanisation. Nevertheless, even though the Lake District is within commuting distance to Galway and other centres near the region, unemployment levels are among the highest in the country. The region has fallen behind average Irish living standards, and the inhabitants press for better roads and health services. Wages levels are low, which for some industries can be an advantage, helping local industries compete and providing employment within electronics and optical equipment industries, for example.

Purpose of the Lake District Enterprise:

To establish an enterprise network
To promote the Lake District Region nationally and internationally
To promote economic activity in the region by creating employment in agriculture, tourism and industry
To try to reverse the trend of population erosion.

(Lake District Enterprise Articles of Association, 1997)

The Lake District Enterprise was founded by eleven people, three women and eight men who constituted the board. They were all living in the area and were actively involved in business and social organisations. The new organisation was launched in 1997 after a preparation period of 18 months. The founders worked with all the communities in the area to see what the communities needed. They identified the level of interest in the initiative and tried to encourage others to take part in it. During the preparatory period a group consisting of community and development agents and other interested parties was also formed, and held public meetings throughout the area. It was considered a major event when 700 people came to the public launch held in the community school during the summer of 1997. A manager of LDE said to us: 'There was huge excitement at that time. We were going to do something for ourselves.'

This led to the establishment of an organisation with a broad span of development aims, although with a special focus on industrial development in an effort to change the trend in a decline of population. The Lake District Enterprise covers a range of activities from enterprise start-up support, training programmes (especially in computer skills and customer service), tourism (information and marketing) and cultural events (caravan festival, hill walking club, angling). The new relations between the tourism industry and non-business organisations such as traditional music and hill walking are of a special innovative kind. High levels of interest with increasing numbers of participants and a rise in tourists visiting indicates that the initiative may be successful in this regard. The Enterprise has also been active in establishing a Family Care Centre.

The Lake District Enterprise is open for membership from business and individuals, and at its peak there were 400 members, although at the time of research there were about 100 members. The activities run on project funding from private and public entities, membership contributions, sponsorships and general fund-raising activities, such as sports competitions and church gate collections. In the first five years of activity, LDE were able to raise over €500,000. Until recently there have been one or two people working in the Lake District Enterprise office. Again at the time of research, there were no paid staff due to financial problems, and the board was working on a strategic plan to refocus the activities and raise new funding for the enterprise.

The Lake District Enterprise initiative has made itself visible to a greater extent than other organisations such as South West Mayo Development Company (Hannon et al. 2003). There are examples of companies who have relocated to the area or changed plans to leave. There are also companies using the Lake District logo or referring to the lakes in their names, especially within leisure and tourism. Festivals and leisure clubs (such as hill walking and angling) are now established as regular activities, and a number of people have sat the national Internet and Computer Training exams or accessed business start-up help.

Finding and Comparing Regional Identities

There are some broad contextual differences that may explain some of the structural differences between these two regional initiatives. Among these are the systems of political administration. While local democracy is strong and active in Norway, it is a good deal less powerful in Ireland. Ireland is a much more centralised country as far as policy on public services is concerned. Industrial development is related to private interests and national policy. The Lake District Enterprise is therefore in line the national system, with those launching the initiative comprising of private citizens largely relying on private or charitable funding. Another structural difference is the membership of the European Union, with Ireland as a member and Norway only an associated member state. As a consequence Ireland is more familiar with rural development initiatives being organised in regional agencies, such as LEADER for example. A third difference is the funding structure, which is partly a consequence of the differences in EU membership. More than 50 per cent of the Lake District Enterprise budget comes from regional members, the rest from external funds, while the Mountain Region Council is 100 per cent publicly funded (combining different public levels and funds).

Given these caveats, however, we can still ask about the regional identities of the two study regions. We understand regional identity to include the cultural and social representation of a region, to which people relate. Of interest are the traditions and perceptions related to a geographical area both by inhabitants and visitors (Paasi 1986; Lysgård 2001). Relevant to our study are regional identities shaped by historical interrelations and changes in the community. Here we will present and compare the main elements of the two regional identities, and compare them. They include both the subjective representations of the regions (images) and the practical relations (content). Of special interest in our study are the relations between the historical and socially rooted identities and the symbols and activities initiated by the new rural development organisations.

The Irish Lake District and Norwegian Mountain Region also have certain characteristics, history and challenges in common. Housing, cultural traditions, and agriculture are all linked to economic life in both regions. In the Irish Lake District stories about times of poverty such as the great famine (1845–1848) and large numbers of out migration are still told and remembered among people in the region. Similarly, in the Mountain Region the years of mining and marginal farming are remembered and celebrated in festivals and museums, and especially in the UNESCO conservation of 'The Old Mining Town of Røros'. In both regions these histories are reflected in the cultural life of today, and are examples of how sharing and repeating histories and narratives creates regional identities.

For the inhabitants there are positive aspects of living in their respective regions that indicate strong place attachments. The content of this regional identity is made explicit when people describe the basis of their choice of place to live, drawing on features that are shared by inhabitants in the Lake District and the Mountain Region. The surrounding environment and the traditional rural lifestyle are among the most frequently mentioned aspects and affect the overall perception of quality of life. Scenery, clean air and outdoor life are highly appreciated qualities which are also used to attract visitors and potential in-migrants in both places (Hannon et al. 2002; Rye and Winge 2002b). These qualities are clearly important elements to the regional identities. Not only are they shared within the regions, but they act as public goods in Cecchi's terms (Chapter 3), being used for the motivation of community development and the maintenance of the decentralised housing and service pattern.

So, as the new regional initiatives for rural development are established, how do they relate to the symbolic and acted-out regional identities? The most obvious symbols are the logos chosen as headings on letters, web pages and other relevant occasions. We find that development organisations are using typical landscapes, the lakes, mountains and trees, in their logos. The symbols and colours of the logos are chosen so all inhabitants in the regions can identify with them. The logos create an association with a fresh landscape of water and green hills of the Lake District, and a more rough and less generous nature in the Mountain Region. In addition, the logos communicate to visitors and outsiders. By illustrating these central elements of the natural qualities appreciated by inhabitants in the regions, the logos are symbolising important elements of the regional identities in Ireland's Lake District and the Mountain Region in Norway. Nature is used in these cases to appeal to an inclusive sense of identity, thus avoiding overt mention of any of the diverse, or potentially diverse cultural or historical traditions in the regions (see Chapter 7).

An further explicit symbol of the organisations is their name. In both cases names have been invented that were not previously used. In the Lake District, with the intention to involve 26 local communities, and so representing a small part of the western region in Ireland, it seemed natural to let the name reflect the most obvious shared characteristic of the area, the lakes and the rivers.

The Lakes have traditionally been used informally among the inhabitants to represent the area. For marketing purposes a similar name was selected which covered the overlapping geographical area – Joyce Country (named after the famous author), Mountain and Lake District. In Norway, the name of the Mountain Region had no previous history, and came out as a compromise due to the fact that all alternatives had an existing history and were recognised only in certain parts of the new territorial region. The suggestion came partly from the regional newspaper 'Labour Justice' (*Arbeidets Rett*), which already covered most of the municipalities that were to be included in the new regional council. The name 'Mountain Region' was gradually used in more of its articles. It was introduced in the sub-title in 1993, at the same time as the decision was made to appoint a working group to formalise regional cooperation in the Røros area and North East Valley (Nord-Østerdal). In both case regions we find that there was a discussion about the name, and that the participants were eager to find a name free of old or misleading associations.

Even if the regional names are new, there nevertheless exists a tradition of relations and to a certain extent cooperation between the inhabitants. In the case of the Mountain Region, the same territorial unit was used for the regional sports association, and similarly for music and theatre organisations. Even after the re-organisation, competitions and festivals have been cross-county events, and are active arenas for social relations in the area. In Ireland's Lake District sport, music and language play an important role in cultural identity. In addition to the Gaelic Athletic Association, music, dance, drama and sport competitions form the basis of significant and high-profile regional networks in the area.

Regional Identities and Rural Development

Regardless of narratives and images, regional identities cannot exist if the content is not relevant for the inhabitants. We now need to turn to this question. The festivals and sport competitions and the maintenance of historical and cultural cooperation can be taken to indicate that a regional identity does exist among people, in that they recognise the region as a significant entity that they have a stake in. The quality of life shared among the inhabitants indicates that regional identity has positive aspects to it. But this is a general characteristic of rural areas in the two countries of study. Does this pre-existing identity cover the new territorial units invented by the rural development initiatives? And does it involve the rural development challenges that the initiatives are facing today?

Both the Lake District and the Mountain Region are experiencing social changes which are affecting regional identity. Everyday ways of life are gradually changing, especially regarding work and services. As for services, the general development towards specialisation and increased consumption among the inhabitants has strengthened the role of the regional centres. The market towns of Ballinrobe, Røros and Tynset have thus become arenas for a more regional cooperation. Similarly, economic life has been restructured towards fewer and larger companies and production plants, and the extent of commuting to places of work outside the local communities is increasing. All in all, as a result of such social changes over the last few decades, regional interrelations are becoming much more normalised daily practice within working life and service provision, in addition to the cultural arenas and activities that previously common took place on a regional level. How are these social changes affecting regional identity? And is there a consciousness among people of these changes?

In the Mountain Region there are mixed feelings about regionalisation. On one hand, people appreciate the extended scope of public services and career opportunities but, on the other hand, daily (or weekly) commuting, merging of primary schools and loss of local shops are threatening the rural lifestyle of local communities and dense neighbourhoods. Some people think of regional cooperation as centralisation and therefore a threat to what rural development should be about. This links to an idea of development as maintaining the way of life of an area rather than necessarily having to change it. Such scepticism and resistance was expressed amongst local community groups in Ireland's Lake District in the preparatory phase of their project. In the rural Mountain Region of Norway there was heavy resistance, and discussion in the newspaper, to the idea of inter-municipal public services when the issue were first raised.

These experiences indicate a difference in opinion in both case study regions, where individual decisions and practical proceedings do not necessarily correspond to more idealistic principles or the general trends of social change. Considering the relations between principles and practices in such identities, Lysgård writes: 'By analysing social practice and processes attached to these themes, it becomes possible to understand if the region and regional identity in question actually has any substance as institutionalised practices and/or a part of peoples consciousness.' (Lysgård 2001, 260). Lysgård studied initiatives for inter-Nordic cooperation, and found that as long as there is no practical or experiential content, people often become indifferent to the common region (Lysgård 2001). In the Irish and Norwegian case we might similarly suggest that the regional identities of the Mountain Region and the Lake District are not strong, and do not yet have much substance in peoples consciousness and practices.

A final 'test' for the strength and content of regional identity is suggested by Paasi (1996), who suggests that strong regional identities are often confirmed and institutionalised in language, defined borders, and organisations. Neither in Ireland's Lake District nor the Mountain Region are these criteria met, and indeed, we might not expect such embeddedness within local culture of relatively new social institutions. But such thinking shows the difficulty of creating new social forms on the regional level. For example, there exists no collective name for people from the two case study regions while in Norway people from the

North East Valley are *nordøsterdøl*, for example, or from Trondelag, *trønder* and these are in daily use among Norwegians.

In Norway the Mountain Region Council have actively initiated at least two new institutions, neither of which advertise the regional organisation to any great extent. One is the decentralised service of university courses, which bears the name of the universities offering the courses. As a result, it is vague about being a regional university and initiative. The other is the active lobbying network the Mountain Region Alliance, which includes four neighbouring regional boards. Both are examples of success of regional cooperation, but neither of them explicitly represent the Mountain Region identity. An interesting exception, however, is the new association for organic farmers, named 'Food from the Mountain Region'. The new organisation has members from the majority of municipalities, but the Mountain Region Council itself has not been involved. Within tourism, which is considered a potential source of benefit for the regional objective, the situation is different. The existing Røros tourism board have rejected the idea of changing their focus of promotion from the image of 'The Old Mining Town of Røros' and the issue has not been further developed.

The Irish case is somewhat different. The Lake District Enterprise have implemented some new regional organisations, which cover the region and bear names that include the 'Lake District', such as the Lake District hill walking club and the Lake District Tourist Office. Active participation in the fishermen's dispute (see Chapter 7) helped to highlight the common interests among inhabitants in the region. In addition, there are a few new business initiatives established with the aid of Lake District Enterprise that use the name and logo of the region.

As for formal territorial borders the two regional units also differ. In the Norwegian case there are clear borders according to the participating municipalities Still, however, there has been some fluidity as two municipalities have decided to resign while one has joined the 'region' since it was established. The Mountain Region concept is, in any case, not very explicit in the area and appeared in our research to be outwith the everyday consciousness of the inhabitants. In the Irish case the inhabitants seemed to actively relate to the concept of Lake District, even if the geographical boarders not precisely defined. The individual, company and voluntary group memberships have a certain amount of geographic dispersal through the wider area. In this case we might, in contrast to Paasi, see this as a positive and inclusive process of identity formation, with few traces of stringent social demarcation between 'us' and 'them' that the clearly defined boundaries of the Norwegian Mountain Region brings.

Altogether we find some strong and common cultural and social qualities characterising the respective regions. However in both cases there is only a relatively weak consciousness of regional identity. Consciousness in regard to on the one hand local and community identities and on the other to broader linguistic or cultural identities seems to be much stronger. We will continue by comparing in more detail the actual activities of the new rural development initiatives in the process of new regional units.

Comparing Regional Identity-building Activities

The regional identities promoted by the Lake District Enterprise and the Mountain Region Council in Norway have shown to be surprisingly similar, both in content and symbolically. However there are striking differences in the way the initiatives were initiated and are operating. These variations are related to the process of building regional identity. The lines of action in the two initiatives are different, and regional identity-building have been more crucial for, and a more explicit element in, the strategy of Lake District Enterprise compared to the Mountain Region Council. Here we will further explore the variations in regional identity-building activities in the two case initiatives.

The Mountain Region Council and the Lake District Enterprise are strongly related to the 'local community development' tradition which has been developing since the 1970s both in Europe and elsewhere around the world. In brief, such initiatives are characterised by local development activists, broad participation, common challenges and local resources (Farner 1988). The ideology of 'empowerment' is an important element, considering consciousness raising of the individual and mobilisation of the community (Rowlands 1995; Nelson and Wright 1995). Other terms such as 'counter development' (Galjart 1980) and 'counter-force' (Wollan 1994) highlight the motive behind reactions to changes and challenges of a national and global origin. Although the Irish and the Norwegian cases in our study are of a regional kind, the purpose and type of activities are similar to local community initiatives.

Previous studies within the local community development tradition have rarely dealt with identity-building. However, the strategy and ideology of such development, focusing on human resources, participation and shared interests, have an obvious function in the mobilisation and creation of social capital. Here we will transfer the analysis of community development to a regional level and explore some analytical themes regarding the building of regional identity in the Irish and the Norwegian examples.

First, the theme of common challenges and local resources. As demonstrated in the previous section, these criteria seem to incorporated within the case study initiatives through building on existing territorial identities and qualities. Both the Norwegian and the Irish initiatives have been mapping shared interests in the region, the Lake District Enterprise by preparatory public meetings and cooperation with local community groups, and the Mountain Region Council by coordinating challenges in the member municipalities. Both organisations have also chosen activities according to resources and traditions in the area. One example is the newly established hill walking club in the Lake District which is a further development of a traditional activity in the area, organised to make the outdoor environment accessible to more people. Also the new music festivals are modern varieties of traditional activities that appeal to new participants. The culture and leisure-based activities are a further part of the efforts to develop of a green tourism industry led by the Lake District Enterprise, which has chosen to engage with the local traditions and qualities appreciated by the inhabitants. Human resources are also being developed to meet new challenges, through popular courses in computer training (Lake District) and decentralised university education (Mountain Region).

A second theme relates to the development activists who inevitably become closely associated with the initiatives themselves. For both case studies the activists are characterised as part of the elite in the communities. In Ireland's Lake District the activists were well known as businessmen and women and community entrepreneurs. Similarly, the municipal mayors who are elected leaders of the communities led the Mountain Region initiative. It has been noted around the world that development initiatives are normally led by persons from the elite, and this is often reflected in the community structure (Almås 1985). In both these case study regions in Norway and Ireland the activists are people with relatively high or important positions, either within business or politics, and generally considered to be successful by other inhabitants. Conversely, this can lead to perceptions that such people only carry out such work for their own, demonstrable, advantage. Further discussion of leadership in rural development projects can be found in Chapter 6.

The crucial question for a bottom-up development is whether other inhabitants identify with those who take the initiative at the start. In the Mountain Region Council, mayors are elected for a four-year period via the local democratic system to represent the interests of the inhabitants in each municipality. This democratic basis should guarantee a degree of support among the inhabitants in all participating municipalities. In Ireland's Lake District the activists were all from Ballinrobe, but wanted to involve actors in a larger area. One of the directors told us 'We didn't want Lake District Enterprise to be perceived as an exclusively Ballinrobe-focused activity.' Therefore, an extensive preparatory process was carried out involving actors from all over the region. The strategies in both cases were intended to relate to and eventually be positive for regional identity-building. However, experiences did not turn out to be entirely easy.

In the Norwegian case we have the example of two municipalities resigning from the Mountain Region Council. Out of a total of nine municipality members from the start, only six have been members from the start. This might indicate a varying strength in identification with the region among the members, especially between the fringe and the core members. It is also a reminder of the strength of more localised identities and democracy that implies a limited regional solidarity. In Ireland's Lake District in 2003, as the Lake District Enterprise was short of project funding and had to readjust its strategy, the question was raised whether or not the Ballinrobe origin of the activists had been a drawback for regional identity after all. These experiences of course relate to the lack of identification with the region from the beginning, but also to the difficulties that are faced in these kind of linked social and economic development initiatives. Certainly there was resistance among local community groups in the preparatory phase of Lake District Enterprise. The community groups felt constrained as rural development actors, and even if they were in agreement that the Lake District Enterprise would only supplement other initiatives, the resistance was not fully overcome. In any case, these experiences indicate that there are more relevant elements than just the initiative for a common identity. But attitudes can change over time, especially as results from the initiatives become visible.

A third theme of identity-building is participation in the initiatives themselves. Participation in planning and implementing new activities create social relations and shared identities. How do people actually participate in the new projects?

Lake District Enterprise is a private initiative, and the members relate directly to the new initiative. As participants at courses and festivals or contributors in fund raising, members, other inhabitants and visitors have been active in rural development activities that might be seen as building an identity for the Lake District. As a bottom-up initiative, local support and enthusiasm had to be engendered before the initiative could be put into real effect. Particularly in the preparatory phase, local community meetings and the many people attending the launch ceremony can be considered as an active mobilisation for the regional initiative and, therefore, as positive identity-building. This is in contrast to the Norwegian case where it is primarily the local politicians and municipality head administrators who are involved in the Mountain Region Council. Only a few inhabitants participate in the projects run by the organisation. For inhabitants in this region the only way to be involved in planning and policy for the regional initiative is indirectly via the local democracy.

If, as we have outlined in this chapter, identity exists primarily through social relations (as evidenced through various symbols and practical activities), we expect direct participation to be most effective for regional identity-building and, therefore, the structural differences have implications for regional identity-building in the two cases. For the Lake District Enterprise there was an awareness of the need to develop a Lake District identity in order to implement the regional initiative, establish a new territorial unit, and even to finance the activities by local funding. As for the participation criteria, the Lake District Enterprise strategy makes it the more active regional identity-building initiative of the two.

Regional identity has to be verified and renewed over time, and social relations kept active and narratives repeatedly shared to be alive. What were the experiences in Ireland's Lake District and the Mountain Region in Norway after six and eight years of practice? The enthusiasm for Lake District Enterprise seems to have somewhat calmed, and while the Enterprise is inactive the regional identity-building effect is unclear. The Mountain Region identity is still vague, but over the years there is gradually more support for it, especially insofar as it is able to contribute to a high quality employment, housing and public service region for its inhabitants. This indicates a gradual growth in strength and consciousness of the Mountain Region identity, while the Lake District identity has proved to be vulnerable and dependent on continual new initiatives to develop further.

At this stage, with experience gained, it is the activities and experienced results from common efforts that determine the success of the regional identities. This can be considered as a fourth theme in our comparative analyse of identity-building in the two projects. It reminds us of the interrelation between the image and the content of the regional relations, or in other words, the fruitful cooperation as a positive aspect of identity development. We can trace this theme by considering the networks that were established and operate in relation to the rural development initiatives. Who is participating and what are the results of the efforts?

The Lake District Enterprise projects have involved both business people and other inhabitants. Activities covering a broad spectrum from business supervision, the organisation of outdoor and cultural leisure activities, to family care services have involved a broad span of participants. These have all been understood to be relatively successful projects, which seem to have contributed to a developing regional identity among their participants. Further evidence of an emerging regional identity is that some of the participants have started businesses that would otherwise not have been carried out, and 'foregone maximum profit for the sake of area development', as one activitist put it. The Mountain Region Council projects have first and foremost involved politicians and local public administrators in issues such as taxes, transport services and community planning. Most inhabitants relate to these efforts only via the news. However, a few inhabitants are directly involved, by making use of decentralised courses in higher education or training at their work place. The Mountain Region Council has concentrated on a few issues and operates on a long-term time perspective. So far this has been reasonably successful, and there is no doubt that the efforts have made a difference for community development in the region. In a feedback loop, experiencing the results of positive cooperation has had its own influence on building a regional identity – although the exception is the two municipalities that resigned from the Council.

It is a little more difficult to explain the situation of Ireland's Lake District, in which sustained backing and continuity for the initiative has been hard to come by. There have been few direct failures among the implemented projects. One problem came, however, in the plans for a new hotel that were announced at the launch of the new community enterprise. There were high expectations for the hotel to be a driving force for more tourism and it was a big disappointment when the plans were shelved. This might have wounded the still-emerging sense regional identity. To some extent the planned hotel had become a key symbol for the regional initiative, and its cancellation became a symbolic defeat for the idea of regional cooperation on rural development.

Identity-building also has external extensions. Ireland's Lake District is recognised by visitors, both nationally and internationally. But the Mountain Region is also recognised among national politicians and planners in Norway. The effect of national lobbying has contributed to the regional identity of the Mountain Region, although not convincingly enough to put an end to the inter-regional competition, such as the ongoing discussions about localisation of new health services and police offices, for example. It is still to be seen if the external tourism and business efforts will strengthen the Lake District identity and interest for further regional cooperation, and whether the lack of links to local democracy can be overcome by following a different kind of model.

The rural development initiatives have experienced both internal enthusiasm and resistance regarding suggested activities, which indicates that some types of community change are more welcome than others among the inhabitants. For instance, new business initiatives may be supported, but only under certain conditions. One such condition in rural communities includes traditional rights concerning access to natural resources. These are strong marks of social and cultural identity (Forberg 2002), even if the importance for income and the number of people actively using the right is decreasing. This was proved in the Lake District in the fishermen's dispute (see Chapter 7). The same phenomena are illustrated in a Norwegian study of mountain hamlet identities: 'Urbanites are wrong if they think people living in the mountain areas do not eat kebab. But if you touch the grazing right there will be trouble' (Forberg 2002).

Comparing regional identity activities in the two initiatives reveals complex processes within local and regional society. The relationship between these social processes and development outcomes is inevitably hard to predict, and indeed not straightforward to discern in retrospect. We can draw a negative conclusion that even if there is a consciousness about the importance of a shared territorial identity, and even if the strategy includes broad participation and a mapping of interests, there is no immediate certainty that a strong regional identity will result. But there are other aspects that might be taken on more positively in the analysis, and it to these that we now turn.

Understanding External and Internal Problems in Building Regional Identities

When we chose to study regional identity-building as a part of two rural development initiatives, it was to explore the potential for enabling such identity to become a form of social capital. It could also be both a proactive and reactive force in common efforts for community development. But there are a number of aspects that emerge from these case studies that complicate the situation.

First we must consider the possibility that regional identity cannot be seen as merely a scaled-up version of a person's local identity. In the opinion of

Salomonsson, regional identity only matters as a 'costume' when operating outside the region, to indicate a difference. Within the region local identities matter more (Salomonsson 1996). Regarding rural development initiatives, this indicates that regional identity-building is most relevant in strategies of communication and negotiations with both the surroundings and external actors. This principle might explain the experiences of the Mountain Region Council in succeeding relatively well in external matters, such as lobbying for taxes, transport and communications, without being able to deal with internal coordination such as putting an end to negative competition between municipalities. Similarly, in the Lake District there has been success in promoting the region as a tourist and recreation destination, while at the same time internally there is a discussion whether Ballinrobe is favoured by the Lake District Enterprise.

Secondly we must be aware of the distinction between regions of a functional kind and regions as a cultural unit. Functional regions include territorial units defined by administrative borders, industrial connections and so on, while cultural regions are characterised by common history, language, religion, identity and other longstanding relations. These different processes operate simultaneously, but while functional regions are a result of external forces, cultural regions are constructed internally (Salomonsson 1996).

The challenge of globalisation that the Irish and Norwegian case regions face is typically an external force. The new regional units established to cooperate in developing industries and communications, housing and public services for the inhabitants are primarily of a functional kind. Therefore the attempts to build regional identities (with a 'cultural' slant to them) within the framework of a functional region seem paradoxical at a first glance.

There are also internal forces contributing to the process. The initial actors are local and refer to historical and cultural relations and traditions dating back many years, which are important for a common territorial identity. We might therefore think about Ireland's Lake District and the Mountain Region in Norway both as functional regions and as cultural regions. The external forces may articulate with internal cultural processes, and regional identity and other types of social capital are obviously important elements in strategies for endogenous ('functional') rural development in the case study regions. However, inevitably functional and cultural regions do not operate according to the same kind of territorial boundaries, and do not usually develop at the same pace or even in the same direction.

This leads to a third aspect for consideration, which is the time scale of development processes. Even if the inhabitants gradually change their daily practice, becoming involved in more regional relations (such as commuting for work, schools and entertainment) they will uphold cultural identities and values inherited from previous generations. There is a time lag between changes in functional regions and cultural regions, between for example, the establishment of a regional family care centre (as in the Lake District) or a decentralised university (as in the Mountain Region) before the inhabitants consider it as a part of a common consciousness or identity. Even the frequent use of the name 'Mountain Region' in the regional newspaper, or church-gate fundraising in the Lake District, are not likely in themselves to significantly speed up the process of building regional identity. This may explain how it is possible for the two case study initiatives to achieve a measure of success in community development without a great effect on the building of regional identity. The improvements largely contribute to a functional type of region, which can change relatively fast, while regional identity, relating more to the cultural aspects of a region, requires more time for the new social relations to become strong and conscious. The contribution of both to positive and sustainable development is clear.

A fourth circumstance is the underlying values or 'strong judgements' that are embedded in existing identities, which explains why cultural identities offer resistance when challenged (Taylor 1964; Forberg 2002). In the common regional initiatives these have to be considered in order to activate the social capital in the new regional identities. In both cases we find a strong but generalised attachment to the territorial area among the inhabitants. Even in a situation of the centralisation of housing, employment and public services in both countries, inhabitants of Ireland's Lake District and the Mountain Region of Norway have tended to maintain their population compared to other rural areas. Even as traditional industries decline, thereby challenging the economic basis of the communities, alternatives are developing in both regions. In the Lake District, where unemployment is high, the ability to engage in self- and multi-employment is developing into new industries, both in tourism and a variety of other services. In the Mountain Region new employment opportunities within public and private services have more than replaced lost jobs within manufacturing industries and agriculture. In both cases an increasing number of people are commuting to work in centres within and outside the regions. The willingness to change daily practice in order to remain in the area indicates the strength of territorial identities, either regional or local. This is the internal force that the Lake District Enterprise in particular have been mobilising and, to some degree, also the Mountain Region Council.

These examples illustrate that the success of new initiatives depends on how far they match the valued qualities and underlying sources of identity in the communities. In fact both initiatives have experienced the need to adjust their strategies in accordance to internal responses in order to continue. The Mountain Region Council has, as a result of an evaluation in the early 2000s, ended the strategy of regionalising public services in favour of coordination between municipalities of investments and planning for housing and employment. The need for Lake District Enterprise to reconsider its strategy, probably towards a focus on internal support, is part of the same process. The new arenas that are created will undoubtedly have to repeat the exercise of adjustment and readjustment again in the future. This is necessary to deal with the dynamics of a continually changing community situation.

The need for time and continual adjustment reminds us that regional identity-building is a long-term process. Relevance, interest or consciousness of the content of the regional relations varies over time. Depending on changes in identification, regions may go through a 'life cycle' from emerging, to existing and then finally disappearing as socio-spatial communities (Paasi 1996). Both regions in our study are new constructions and are at an early stage in their life as identity regions. If further cooperative activities and initiatives are beneficial and achieve relatively wide participation or interest, the regional identity is likely to grow stronger.

At this stage, the Irish and the Norwegian regional identities can, therefore, only be considered as supplementary to the local and national identity. However the regional initiatives have engendered social capital that otherwise would not have been mobilised. We might therefore comprehend regional identity-building as the challenge of making the most of existing territorial identities as social capital for common rural development. This is highlighted in Paasi's theory about the institutionalisation of regions (Paasi 1986, 1996). New organisations can contribute to making the regionalisation process explicit by implementing formal organisations, giving the territory a name, and setting an agenda of common challenges and shared goals. The new arenas for regional development are a framework for coordinating initiatives and resources, and a channel for exploring national and international influences to make the 'regional voice' heard.

Conclusion

The Lake District Enterprise and the Mountain Region Council face similar challenges of rural development, and both have established a new territorial unit for regional initiatives. By comparing their varying strategies and experienced results we are able to present some conclusions about regional identity-building in rural development initiatives.

- The regional identity-building activities we have studied refine qualities, history, traditions of an area into a new name and logo, and communicate the image through promotion for tourism, coordination of services and lobbying of national political structures.
- Regional identity-building is a long-term process. It takes decades of active relations and representation (image or visibility) building to create a strong and conscious regional identity.
- Regional identity varies among actors, dependent on their involvement, and the relevance and results of the initiatives.
- The content of the new territorial units are primarily of a functional kind, and regional identities must be considered as supplementary to the strong existing local identities.

- The success of regional initiatives depends on the matching of shared underlying values of local identities.
- Rural development initiatives institutionalise new territorial units, both as functional and identity regions, by setting common rural development challenges on the agenda, presenting regional cooperation as a strategy, and creating arenas for discussion, networking and experimenting.
- The new regional initiatives we studied have engendered social capital that otherwise would not have existed or been directed into community development.

The Mountain Region Council and the Lake District Enterprise are examples of active responses to global challenges through cooperation for rural development on a regional basis. Similar to previous initiatives they are experiencing the dilemma of balancing available time and capacity with high expectations for quick results. As such they must also consider the plurality of local identities and eventually readjust their strategies and meet the challenges of changing social practices.

Chapter 6

The Role of Identity in Contemporary Rural Development Processes

Frances Hannon and Chris Curtin

Introduction

In this chapter we present some aspects of the relationships between identity and the environment, culture and the leadership of development in this book's case study areas (see Chapter 1 for introductions). We maintain the overall contention that some areas appear to be adapting faster and more successfully than others to the challenges posed by the shift to a market economy, mobilising themselves to defend their position better in global markets. In paying attention to the nature of personal, collective and territorial identities, we explore how the variety of affiliations that people may claim affects how they take part in rural development initiatives and in processes of rural restructuring more generally. This introduction outlines our interest in identity as a topic in rural development, and draws attention to the theorisation of identity and development as an important step in understanding case study situations.

Rural populations are attempting to secure a livelihood by diversifying their agricultural activities and engaging in new ones such as tourism. At the same time people are moving into rural areas looking for a better quality of life and these in-movers often have different perspectives and ideas on how local sustainable development should be achieved. The emergence of new economies in rural areas can challenge old identities and, in the new valorisation of local resources, there are new opportunities for local development. It could be argued that current rural restructuring processes provide a potential opportunity for the construction of new identities, or the strengthening of existing identities. Ray (1999), for example, considers that the forging or strengthening of identity, utilising primarily cultural, historical and physical territorial resources, is central to the territorial approach to rural development. Attempts to construct new rural identities can be a stimulus to mobilise the collaboration of various actors in new development processes. We can thus examine the extent to which identity is a creative force that is capable of being harnessed in the service of rural regeneration.

Ray produced a number of papers that outline how a sense of collective identity can be used in endogenous rural development (Ray 1998; 1999; 2000). Although his empirical basis was a study of EU LEADER projects in the west of Scotland and northwest France, his main contribution is a theorisation of the relationship between identity and rural development in contemporary rural Europe. Ray (1999) argues that it is possible to equate marginalised territories with vulnerable peoples so that crises in regional or sub-regional development mirror identity crises in peoples and their governance. An aspect of these crises is what he understands as a human need to belong to a place and to be able to make a contribution to the ongoing development of that place. Ray draws a parallel between 'the territory's need for endogenous development and the psychological need of individuals for a sense of identity' (Ray 1998, 14). Individual and collective identities can be seen to have a spatial dimension and the construction of identity involves developing a relationship with place. Thus, the argument goes, the formation of multiple social and territorial identities can be seen not only as a factor in development processes but also as reflecting the need of people to belong and to collaborate with others in meaningful ways.

Ray further suggests that strategies for endogenous rural development put the participation of local people, with their knowledge, skills, entrepreneurship and commitment, at the centre of development efforts and, in these participative processes, 'people and their self/social identities are both the raw material of the development process as well as the goal of that process' (Ray 2000, 452). He suggests that by marketing a sense of local identity through 'niche products' and tourism services, the identity will be taken on by people living in the area, who will then be more empowered and able to participate in rural development (Ray 1998). We can see here how identity could be affected by human, cultural and social capital in the ability to manufacture and market niche products, and in the extent to which the sense of identity constructed therein is transmitted to other parts of the population. This is also an important element of the productive capacity of networks.

At the same time, views of what constitutes identity can be invoked in over-simplified and manipulative ways, and thus contribute to injustice and strife. As Shucksmith (2000, 210) writes:

> It is vital to recognise and manage the very real conflicts of interest which exist within such symbolically-constructed 'communities' or 'culture-territories,' the obscuring of which may contribute to exclusion. The very process of symbolic construction of culture-territories will exclude and disempower some residents of these localities if they do not feel affinity with the constructed cultural identity.

Although both Ray and Shucksmith are referring in the first instance to LEADER projects, the principle of using 'identity' in rural development is becoming more widespread, as the empirical material presented in this chapter shows. The main research question to emerge from this discussion centres on how the identities that people have and claim in rural Europe get mobilised within rural development discourses, and what the positive and negative effects of this process are.

Senses of identity reflect local culture, heritage and the physical attributes of a place as well as individual and collective identities. The vitality that can arise from a sense of place is a reflection of the connections of self with the physical environment and with significant others in that place – family, relatives, friends, colleagues and other residents. Identification with place and the awareness it brings of all that is associated with that place – landscape, tradition, folklore, culture, change and inheritance – also brings with it a sense of community (Canavan 2002). At the same time, one can make a distinction between local cultural identities, formed through everyday living, and the construction of territorial identities as part of a planned development process, or in other words, where collective identity formation is a deliberative act with a particular goal in mind (Hansen 1999). The potential tension between the two can be seen as important in many local rural development processes.

A consideration of the concept of identity and its role in new development processes would thus seem to be timely. Researching identity may prove essential in deepening our understanding of the role of social and human capital in development, highlighting conditions in which people are more likely to engage in participative development. We explore two ways in which identity is commonly asserted in our case study areas: in relation to the environment, or the place where people reside and feel a sense of belonging towards, and secondly in relation to ideas about culture, or the ways of life and heritage that a group of people have. We then change the focus in order to examine the importance of leadership in the development process, showing how identity can affect and be affected by the specific ways in which rural development occurs. Finally, in the discussion, the heterogeneity of forms of identity in rural Europe is described in relation to the variety of development practices are present. We stress the importance of inclusive identities in rural development.

Identity and Environment

This section discusses the importance of the relationships people have with their environments in the formation of their identity. It presents some of the ways that people articulate a sense of place that emerges through activities that they undertake, and how this can provide a basis for participation in rural development. Further investigation of the environment as a factor in rural development can be found in Chapter 7. Here, however, the primary focus is on identity.

The countryside is a distinctive environment from the urban one in spite of the now almost universal provision of basic infrastructure in rural areas. Rural dwellers in the case study areas spoke of features of rural life that they appreciated, which included living 'close to nature' with its accompanying 'space, peace and tranquillity'. At the same time, however, many people drew attention to the active ways in which they were working with and changing the land, and distinguished these from more conservative ideas of a rural idyll. For example, in the island of Skye in Scotland many of the crofters in our case study not view the landscape as static but as something changing through the work of their hands, and this labour links them to all their predecessors who worked the same land. For the crofters, the environment is not just plants, soil and animals. It also relates closely to work, heritage and land tenure and thus is political as well as personal and communal. In the Irish Lake District as well (for an introduction, see Chapter1), tangible relationships with nature colour peoples' experiences and are a source of satisfaction as well as hardship.

> Even though farming is hard, to walk out in the morning to the birds singing, to be on your own land, working it – it's a good life. (Interview, Lake District).

Particular rural activities, 'working with nature', 'creating it', 'transforming it', therefore inform rural identities. Furthermore, even when rural dwellers are not farmers they often have a particular experience of the environment that is part of their identity:

> The landscape – lakes and mountains – they appeal to me. I have a lot of good memories of simple pleasures ... camping, fishing. And the friends, a lot of good people. I think the combination – the landscape and the people, it's very irresistible, it's an attraction. Throughout the years when I lived elsewhere my heart was still here. You know through the mountains I re-engage or I go down to the lake and I feel like I can touch base again, and the good friends as well, there's something very earthy about the place, it's quite simple. Again, as well as all that, there's a lot of layers of history. (Interview, Lake District)

Another man explained:

> My life is steeped in the traditions of the lake. Even if the name of the Lake District wasn't there the lakes are the centre of the region for me. I love the lake. I would be very proud of the lake, fishing, the associated culture and tradition that go back for generations. Building boats is a part of the local fishing tradition and as a child my summers were spent hammering nails into oak and larch. (Interview, Lake District)

This characteristic – the relationship with all that a place embodies – can be harnessed to give an impetus to initiating and collaborating in local development processes.

Place identity is an important resource for collective action as well as for place marketing strategies that build on the uniqueness of place. In the island of Skye in Scotland, identification with the physical place – the feeling of being part of a unique place – is a common identity that can embrace other Skye identities such as Gaelic, local or Highland crofting, and township identities.

Development organisations in recent years have attempted to emphasise and utilise this place identity as a suitable scale at which to operate in development projects, although, as the Scottish case study describes, this is a process which is highly tensioned and may lead to conflict between such different scales.

A common identity, real or potential, based on aspects of what a particular place embodies, also formed the basis for constructing new territorial identities in the Norwegian and Irish study areas. What is also of note is that local culture and sports, which are often features of place identities, are also sources of commonality that can embrace other identities and unite different groups within a community. In Sotkamo in Finland organised sport allows cross-group solidarity to be built up and new networks forged, with different groups learning that they can work profitably together (Chapter 2). Similarly in Leksand and Rättvik in Sweden there are examples of cultural and sports events forming the basis of network formation between business, community and voluntary groups (Chapter 4).

In addition to the physical environment, the quality of life referred to by people living in the study areas included the opportunity to be part of a particular place and community. In the Irish Lake District, some of the quality of life features most commonly cited included, in addition to the natural environment, friendly people, the sense of belonging, safety for children, cultural activities, the appeal of small town living and economic considerations such as cheaper housing. As is well known, this quality of life is attracting people to rural areas and is important for maintaining people in them. Many of the entrepreneurs in Leksand in Sweden, for example, are in-movers, who first came to Leksand as tourists and were attracted by the quality of life it offers.

It is of course very difficult to isolate specific environmental factors in the construction of identity. This is because the sense of place that many people have is articulated through a combination of work activities, social relationships and feelings of belonging that are situated in, but not always directly derived from, a physical environment. But we can recognise the importance of how people represent their environments in their senses of identity. In the next section, we move on to discuss how ideas about culture similarly inform concepts of identity.

Identity and Culture

In the case study areas over recent years, there has been a general increase in interest in cultural heritage as a local resource, both for its own sake and for the development of specialised area products. There is support for the revival of festivals, sports, and traditions in food, handicraft and local skills. In this chapter, we describe how some of these activities can play into rural development discourses.

The Kainuu region in north east Finland has maintained its own dialect as well as its local traditional handicraft and food expertise and this appreciation of the cultural heritage and the range of local resources are now being used in innovative ways. These form the basis for both local and far reaching networks through which the small and 'remote' municipality of Sotkamo is making a contribution to regional and national life. Continuity in the use of old farm-based networks and institutions are rooted in a sense of local identity related primarily to place and land ownership. The forests are an important part of the local economy and a source of recreation, but also a source of 'peace of mind' for local people. Historically, Kainuu was known as the 'hunger land' because of its harsh living conditions, but in this process we can see how cultural activities can result in a new understanding of the environment. In Chapter 2 Lehto and Oksa also describe how the development of winter and summer sports in the environment of Sotkamo have contributed to rural development.

Similarly, in the Scottish study area of Skye and Lochalsh crofting identities and networks characterised by collective action are implicated in frameworks for new local development processes. According to respondents there, crofting has always depended on successful cooperation amongst crofters. In a practical sense crofting has provided remote rural communities with networking tools: a need for cooperative labour, a strong meeting and committee structure, and deeply embedded social relationships. More than this, crofting can be seen to function as a claim to, or a representation of, a valued style of communal or cooperative working. As such, maintaining crofting maintains the symbols of close social relationships. The potential of crofting as a symbol of collaboration for local development is also now being recognised by development agencies operating in Skye and Lochalsh.

Some of the common cultural features of the Swedish study area that were commented on by informants include the importance of identity, in terms of ancestry rooted in a particular place (originating from the Scandinavian land tenure tradition of dividing the homestead among all heirs), a tradition of flexibility in earning a living, and a tradition of local decision-making. These features play a part in the form that development processes take today. The maintenance of traditional culture combined with the flexibility in employment means that the local people have tended to be simultaneously open and innovative while rooted in tradition. Paradoxically, the importance local people ascribed to their traditions helped them to become the Swedish pioneers of tourism, with the tourist industry dating back to the nineteenth century. This area of Sweden is commonly denoted as the most 'Swedish' place of all. The municipality of Leksand has created an image of good quality of life embedded in a rich culture and a beautiful landscape with meadows, hills, forests and lakes. Handicrafts of the area, such as the wooden 'Dala-Horse', have become a symbol for whole of Sweden. It is interesting that these two municipalities have thus used their cultural identity constantly as a resource for development, and could be argued to have combined the promotion of traditional culture with flexibility and innovation (Jonsson et al. 2001).

The Swedish municipalities have over a long period of time actively utilised history and culture as marketable goods as well as for their own sake. Flexibility and innovativeness can be incorporated in order that aspects of cultures and traditions do not become stereotyped and emptied of meaning. There are growing numbers of people who do not agree with the emphasis on conserving traditions and, in this situation, 'traditionalists' reacting excessively to perceived threats local culture may ultimately stultify the traditions they wish to maintain.

While the consensus from study participants generally was that Leksand continues to be dynamic and prosperous, one participant pointed out that the current political elite, at the present time, might not be so tolerant of controversy. She considered that conflicts and dialogue are suppressed in public discourse in Leksand, which, in the long run, will have a negative effect on innovation and flexibility and the maintenance of culture and traditions. Interestingly in Rättvik, which has had an image of being less successful than Leksand, politics may be becoming more open to dialogue and controversy. In recent years Rättvik has also had a higher business set-up rate than Leksand and a higher population growth.

One of the annual festivals in the Norwegian study area is the Røros Winter Market, which celebrates the region's historical traditions. Again, however, the emphasis on conservation of traditions may militate against new development processes in so far as innovation and adaptability are seen to compromise tradition. Identity involves a dialectic between continuity and change and while connections with the past must be recognised, attempts at construction that are based exclusively on continuity can contribute to the 'marginality' of peoples and places (Bassand 1986). Identity is a matter of becoming as well as of being because it is subject to the continuous play of history, culture and power and it therefore belongs to the present and future as much as to the past (Hall 1990). Flora and Flora (1993) considered that the allocation of resources to risk is central to entrepreneurial social infrastructure as well as the willingness to invest collectively and the willingness to invest private capital locally. The readiness of local government in Sotkamo (Finland) to engage in high-risk behaviour and to take failures in their stride contrasts with the more cautious attitude seen in the Mountain Region where the possibility of losing was avoided.

It is notable that different people within a territory can emphasise the importance of very different aspects of local identity and this can engender conflict. In-movers to Skye in Scotland may not have the same sense of historical association with the area as people with long family roots, and yet may be strongly attached to occupational identities, or appreciate keenly the natural scenery. These varying identities come into play in development discourses exemplified in the conflicts and negotiations surrounding a wind farm development in Skye (see Chapter 7).

In the attempt to construct an Irish Lake District, many people did resonate with the concept of a Lake District. They considered that it fitted in well with all they identified with. Others who lived on the periphery of the area, and who had already bought into and used an older 'Joyce Country' identification (a distinct but overlapping area), did not wish to consider this newly proposed concept. Some further people living in areas considered to be sufficiently successful did not see the point of operating at a larger scale.

What is of note is that the majority of local inhabitants in the Lake District who were interviewed spoke about the process of constructing a new territorial identity from two different perspectives. One was whether the new territorial identity was real in terms of its geographical boundaries, historical and cultural roots, and peoples' sense of belonging to it. Secondly, people discussed whether the forging of this new territorial identity was an appropriate tool in creating a stronger development position for all the communities in the area in an increasingly globalised market. Most people voiced the opinion that, objectively, bigger networks with a stronger voice are needed to compete successfully and, from this viewpoint, the strategy is appropriate and consensus is possible. Thus people with strong local identities (where emphasising differences from neighbouring communities is usually part of those identities) and relatively close-knit communities can be pragmatic and ready to forge alliances. They can work at creating a new identity for the sake of development and, in this case, they emphasised areas of commonality while temporarily suppressing differences. While the strongest sense of identity and the densest networks operate at the most local level, building networks across communities to the spatial level required for strategic development can be achieved. However, whether the difficulties inherent in tight local 'bonding' networks are overcome will depend on the presence or absence of other enabling factors in the area.

Borne out in the Irish study area is that consensus around merely a good idea or strategy is insufficient. People also need appropriate contexts in which cooperation can take place, trust built up and, in this case, a new identity adopted. In addition, while local identity can form the basis for forging networks, development outcomes will depend on the resources, skills and power made available through these structures. Vertical networks through which resources can be accessed need to be developed simultaneously alongside the development of horizontal networks between communities. In the Irish case study, expertise in creating effective horizontal and vertical networks, through specific strategies with professional approaches and structures, was limited

The ways that people identify with aspects of their culture cannot be easily modelled and subsequently 'used' in rural development. Many people are concerned to maintain ways of life and traditions that have been present in an area for generations, and yet rural development discourses require flexibility and innovation. Ways to combine visions and use of the past with the needs and demands of the present and the future are required. The ability of local people to control their own development is perhaps of central importance in achieving an appropriate balance in these terms, and it is to this question that we now turn.

Identity, Leadership and Development Processes

Having outlined some of the ways that people identify with their surroundings and cultural heritage, we now describe the different kinds of people who can initiate and lead rural development strategies. We discuss the extent to which people's sense of their own and others' identity are significant in how rural development processes take place. Of interest are categorisations of 'local' and 'outsider' or 'in-mover', and the extent to which local government or businesses can act as leaders in rural development.

Development processes and networks in the study areas were initiated by different actors, including local business people involved in the community, journalists, government, development agencies and many others. In terms of individual catalysts there were in-movers or people who had considerable experience of working elsewhere who played a role in new development processes. They could 'see the wood for the trees', as one person from the Irish case study put it, in seeming to be less constrained by local mores and often very enthusiastic about the identity they had embraced and their involvement in collective action and community life. In-movers have made significant contributions to local community initiatives in the Irish Lake District. Some of these had a connection with the area (local ancestry, relatives in the area or who were born in the area but spent considerable periods of time living and working abroad).

Tourism in Leksand and Rättvik (Sweden) has contributed to the local economy in an indirect way. Many local entrepreneurs are in-movers who came to Leksand first as tourists, attracted by its culture and landscape. As people for whom the environment has increasingly constituted the primary relocation factor, they moved their enterprises there. A further example of a non-local instigation of a development project from Sweden is that of the former opera singer who sought local support in Rättvik for an open air concert venue in an abandoned stone quarry (Chapter 2). While the municipality and local media initially scorned the idea, the singer tapped into an individual national and international network to lend support to the project, and through this network, was able to mobilise resources including EU funding, to follow through and to establish this project. For the opera venue, some of the difficulties in maintaining local voluntary labour were in the context of a perception that the venue is largely for the benefit of non-local people.

In the Irish Lake District, respondents described a number catalysts for the myriad of development initiatives within individual communities. They included significant local events, innovative ideas that captured peoples' imagination and leadership from local people who identified strongly with their area and 'could never live elsewhere', as well as leadership from outsiders. Challenging events could also provide a focus for new constructions of community:

> I can think of examples in this area where a challenge galvanised all sorts of support. The challenge for local development and for politicians is to present people with challenges. This is not happening. They need to come up with challenges but instead they often come up with dictates. If people are presented with an exciting or interesting something or something unusual they can get their act together and community spirit is mobilised. (Interview, Irish Lake District).

While many examples of leadership in individual Lake District communities were found in this study, there are significant differences in the community capacity required for implementing a vision in a single town or village compared to implementing it across ten or 20 such areas to create, in this instance, a new territorial identity.

The attempt to create the concept of an Irish Lake District was led by a group of local business people who were already involved in community activities. They also considered themselves guardians of the place in ensuring environmentally friendly development. One said: 'we want clean industry that will respect what this region is.' The area has shared geographical and cultural features, including its interconnecting lakes and hills and common local heritage. It seemed plausible, therefore, that, while retaining their local identities, communities in the Lake District could come together at this larger spatial reach to strategically plan the development of local resources.

After five years, however, this group's strong identification with place and culture had not yet been translated into a framework capable of supporting an endogenous development process, even though considerable effort has been invested in developing their area. The actors had been heavily involved in their own community, but had not accessed stable horizontal networks with other communities, or vertical networks to access the requisite expertise.

While members of this group cited financial factors as the major reason for the problems encountered, our study revealed a number of other factors. These included a lack of skills (management, organisational and leadership), and a failure to maintain focus on the original vision of establishing a functioning network at the desired spatial reach. Five years on, most people and communities were not involved in networking activities or events at the Lake District scale. The main activity of Lake District Enterprise has been the provision of computer training in various centres in the area and some local business activity in the Ballinrobe area. Despite discussion about revamping the initiative, there would seem to be significant burn-out on the part of key figures at the helm of the venture. Financial difficulties emerged early on and led to the loss of LDE managers and to the diversion of energies into fundraising projects that did not necessarily further the realisation of the group's initial vision.

From here we could identify a weakness in the entire entrepreneurial social infrastructure. Political clientelism is strong in Ireland and initiatives by local development actors are not always welcomed by local and national politicians. Ireland lacks the strong local government structures of many other countries so there are fewer powerful lobbyists for local development within and outside their area. There is insufficient integration and cooperation between official and voluntary structures in all sectors in Ireland, which is combined with *ad hoc* and arbitrary administrative divisions. Whelan (1996) summarised the practical weakness of Irish territories as deriving from 'a lack of fruitful interchange between those emotional territorial affiliations that live as a deep structure under the surface of daily life, and the arbitrary constantly shifting, centrally imposed administrative divisions used by politicians and bureaucrats as management tools' (Whelan 1996, 118). Failed initiatives and passive communities, displaying varying degrees of resignation, compliance and uniformity, can translate into individuals and groups who do not have the requisite vision and skills and cannot acquire them through the requisite support of local development agencies. When examining local development processes attention must be given not only to inter-organisational networks but also to the nature of government and leadership (Warner 2001).

In contrast to Ireland, in Sotkamo, Finland, formal institutions are evolving in a way that allows them to provide the combination of factors that encourages entrepreneurial local development to emerge. Sotkamo in Finland seems to exemplify the positive effects of the presence of a particular combination of factors, giving rise to a vibrant locally based development process. In Sotkamo there are innovative actors, volunteer groups and networks across diverse sectors including education, sport, tourism, research and the use of natural and manufactured local resources. Dense horizontal flexible networks connect different sectors but vertical links to centres of power have also been forged in which the municipality has been pivotal. They have also generated a particular kind of local development dynamic. Local institutions encourage high levels of local agency and there is flexibility and fluidity between different sectors. Municipal leaders have demonstrated single-minded commitment to local resource innovation and have displayed a readiness to take considerable risks, openness to new ideas and flexibility with regards ways of doing things. The municipality has encouraged local entrepreneurship and innovation and has also been quick to undertake innovative projects, later devolving them back to the public arena. Confidence and openness to ideas is combined with maintaining continuity with the past by building on traditions and incorporating them into new projects.

The municipality's role has shifted from public service provider to catalyst of local development initiatives. It actively lobbies and negotiates with the central administration and has established a network of connections between state ministries, political decision-makers at the national level, regional offices and enterprises interested in developing initiatives in Sotkamo. The municipality has been able to provide technical expertise and knowledge to different projects. There has been flexibility in professional roles adopted by individuals with business people taking up positions in local government or vice versa and individuals in education becoming entrepreneurs. Occasional mistakes, conflicts and struggles have not slowed down their readiness to take risks. Sotkamo won national awards as the most creative municipality (2000), the enterprise-friendly municipality (1999) and the best municipal image (1998).

In the Norwegian study area, local newspaper editors played a central role in the attempt to create a new Mountain Region territorial identity (Chapter 5). They gave the impetus to local politicians and people to consider the potential of a territorial strategy. While the institutionalisation of inter-municipal cooperation took place with the creation of the Mountain Region Council (MRC – a political-administrative body to coordinate public services) in the early 1990s, it was the local newspaper editors who begun to employ the name in the public sphere at a regular basis by using the term in their newspaper. They debated particular cooperative projects across municipality borders and chronicled the discussions between politicians and bureaucrats regarding the setting up of the MRC. They began to publish articles dealing with issues in the *region* rather than in individual municipalities. Instead of being the local paper for Røros, the newspaper became 'the paper of the mountain region'. A decade later, some private enterprises and organisations in the area have adopted the term 'Mountain Region'. It is being used in local advertising and has entered everyday speech to some degree.

However, in the Mountain Region a certain caution and resistance to change were noted on the part of both local people and institutions. A leader of a local industrial development department in one of the Mountain Region municipalities in Norway complained that it was hard to change land use because of a lack of entrepreneurs and because it was so difficult to effect change. The reluctance to change may be, in part, linked to the emphasis on conserving 'the past'. Officials in the agricultural department of two of the Mountain Region municipalities considered that there was both a lack of entrepreneurs and the culture that fosters entrepreneurship. One of the officials cited an example of when government and local agencies had some time in the past advocated fur farming, which later proved a failure and led to bankruptcy for some farmers. It suggested a cautiousness and fear of making mistakes, at least on the part of some officials. Some individuals from the Mountain Region also mentioned that the farming organisation, whose aim is to help farmers improve their lot, was very passive in encouraging innovation. Hence, there would appear to be problems in developing innovations on the part of local people, and on the part of relevant development and sectoral agencies in encouraging it.

In the Swedish study area, as in Finland, there were examples of people in local government who played an active role in stimulating innovation and were working in partnership with local enterprises. It is also notable that the municipal government in both Leksand and Rättvik obtain the village community organisations' opinions in different matters before policies are formulated. Thus, there is formal recognition of the importance of this store of tacit knowledge and of the role of these organisations in collective action with regards to community issues and affairs (Jonsson et al. 2001).

Another example of innovation by the municipality leaders in Leksand was based on the use of local knowledge. The tradition of dividing up the homestead in Rättvik and Leksand means that large timber companies are not the main owners of land. Instead there are many small, private owners, including many persons who originate from the area but live elsewhere. The municipality in Leksand used this situation to attract entrepreneurs back to the area. Around 3000 self-employed individuals in Sweden who owned land in Leksand were found by the municipality and, out of the first 300 contacted, 15 moved their business to Leksand. Ancestry rooted in place is very important for people from this area.

The current business advisor in the municipal authority in Leksand, Sweden has organised meetings of entrepreneurs 'to promote networking' and facilitated women in small furniture firms to implement their marketing ideas. Five firms in the furnishing branch began to cooperate and, on the basis of this, an EU-funded project to stimulate wood processing and handicraft firms was initiated. This had 25 firms from the whole of Dalarna cooperating in 2002. Representatives of the group of mainly female entrepreneurs were planning to visit northern Italy with the business advisor to examine how firms there cooperate and cluster. It is of note that this female business advisor considered that women in the local furniture industry were more interested in marketing and in the more innovative aspects of the business (the men were generally more interested in the production aspect of business) and this recalls the findings in the Mountain Region in Norway where more women than men were involved in local innovation. At the same time, while Swedish municipal governments were actively involved in promoting entrepreneurship there was some evidence that they tended to foster local ventures that fitted a particular model and were not always open to more innovative projects or to the development potential of particular cultural traditions and events. For example, while much support is given to developing business around male-dominated ice hockey, less attention is given to the economic potential in other forms of local culture.

Many different types of leadership in development can be potentially seen as appropriate in terms of senses of identity, depending on the particular local context. We suggest that a strong local identity in a development strategy is more likely to result in the participation of people resident in the areas, as was the case in the Finnish and Swedish cases described above. Nonetheless, there are other examples of development strategies that were highly embedded in a local identity and yet have not been so successful, because of broader socio-economic factors. In this context what is of note is the role of local government in obstructing or facilitating local leadership and local innovation. The lesson is that a concern with 'locality' alone is by no means sufficient in itself to secure a positive development outcome. Similarly, the attempts to construct new kinds of identity in Ireland and Norway have both clearly faced significant difficulties, as is brought out in Chapter 5.

Discussion

Environment and local culture have played important roles in forging individual and collective rural identities in the case study areas. Senses of belonging to a particular place are often used as the basis for collective action and for constructing new territories at a more effective spatial level for economic initiatives and marketing.

A sense of belonging may be vital but it can also give rise to controversy over what constitutes appropriate development in a particular area. This highlights how rural identity and rural communities are not homogenous entities, and how individuals and groups identify with places in distinct ways. The manner in which territories should be developed and portrayed will always be contested. This has been illustrated in the opinions of 'traditionalists' versus in-movers in Leksand and Rättvik in Sweden, the crofters versus in-movers in Skye and Lochalsh in Scotland and people living in the peripheries of the Lake District in Ireland or the Mountain Region in Norway, who identified themselves with different territories from those proposed by local actors initiating development processes.

The increasing heterogeneity of rural identities can, however, contribute to innovative development processes. Understanding and accepting the potential developmental role of constructive dialogue will allow heterogeneous and engaged communities to emerge that are tolerant of diversity and capable of using constructive conflict in development. This can ensure that rural identities do not become rigid but can instead adapt to changing contexts. Acceptance of controversy also facilitates the depersonalisation of institutional politics, which, in turn, allows people to concentrate on process. The importance of creating an environment that depersonalises politics allowing the focus to shift to process was glimpsed in the changes taking place in the two municipalities of Leksand and Rättvik.

In addition to potential clashes between individuals and groups with distinct identities, Flora and Flora (1993) outline other features of rural identities such as the density of acquaintanceship and the overlap of roles that can also make people reluctant to tackle controversial issues, because of the danger of jeopardising a larger set of relationships. In Sotkamo in Finland controversy did not hold up the development process. In the moves towards entrepreneurialism, however, there is also a danger that broader social development and inclusivity may be less emphasised. In Ireland development discourses are often still framed in dichotomous terms, making the chance of constructive dialogue more remote though not impossible.

Ray, in his portrayal of the possibilities of the construction of new territorial identities (as described in the introduction to this chapter) may not sufficiently address the implications of rural territories becoming increasingly heterogeneous in terms of culture, lifestyles and identities. Rather than attempting to develop an area based exclusively on cultural homogeneity, an exchange of ideas, collaboration and the sharing of different cultural traditions across groups may be a more realistic goal. At the same time, increasing consciousness of local history and local heritage can stimulate commitment and sense of belonging to a territory. This can form the basis of creating a community that acknowledges the reality of the past while being open to the possibilities of the present and the future. Thus, while there is increasing migration of people and more diverse rural populations, the best way forward does not necessarily imply ignoring the past history of places and peoples. In the study area in Sweden there has been a long history of linking the past to the present in flexible ways that served the area well in terms of economic development but again there is the ongoing need to blend the old and the new in relevant contemporary ways. What is of note in these cultural processes is that there is openness and flexibility as regards the past and the future and it is this that inspires and stimulates innovation.

Common ground for collaboration in local development and for the emergence of more diverse rural networks can be found in local projects of universal appeal. In Finland the former municipality-driven tourism sector cooperation was replaced by a network between tourism enterprises, the sports college, the baseball team and the municipality, which cooperate and jointly market in a more effective way with tourist marketing being provided by the local baseball team. A millennium project in the Irish Lake District – building a 'medieval' stone labyrinth – unearthed a local tradition of stone masonry skills and led to a considerable new voluntary commitment of time and work. The project led to an awareness of skills and to a demand for stonemasonry work and skills previously not utilised or sought in the area. When common action and exchange did take place, individuals and groups experienced a range of diverse opinions, experiences, skills, new cultural elements and identities.

While cultural activities can have an important enabling effect for local development processes by deepening peoples' awareness of personal, collective and place potential and increasing their sense of local identity, other elements are also required. For example, access to human capital fosters identities more likely to provide the requisite leadership skills for community development. In the case study areas of Sweden and Finland there are many local educational institutions that address local needs. Near to Sotkamo in Finland a number of tertiary level institutions offer educational programmes relevant to the needs of the area and research programmes have been developed around a range of local resources. Such local educational institutes may well have had a significant impact on local identities, human and social capital, innovativeness and creativity. Investment in human capital such as in adult and vocational education and enterprise and information technology training is also likely to generate certain forms of social capital, as described by Shortall and Shucksmith (1998) in their call for training of professional rural development animators.

Thus local collective action and community participation are limited by factors both within and external to individuals and groups. In some of the study areas, such as the Irish Lake District, people had a strong sense of identity, but formal structures did not allow them to be active local development agents. In the Lake District, it was notable that individuals who actively engaged in promoting a wide variety of cultural activities and initiatives were particularly passionate about the development of their place and the potential transformative role of a vibrant culture in local development. Formal institutions must also recognise and encourage this kind of development. Passive communities, displaying varying degrees of resignation, compliance and uniformity, often translate into individuals and groups who cannot obtain the requisite support to be agents of local development and do not regard themselves as actors or agents of development. While complying with local ways and customs may be a manifestation of solidarity, non-participation in shaping ones' community because of potential conflict is opting out of active agency. It is important to recognise that this is not a weakness of identity, but instead is more likely to reflect the disadvantaged socio-economic circumstances of many remote rural areas.

To conclude, people's senses of identity played an important role in local development processes in the study areas, and a situated sense of identity – a sense of place – often facilitated commitment to local collective action. Personal identity contributes to the emergence of leadership in that it leads individuals to develop a vision and to attempt to implement it. In this way it can contribute to making people architects of substantial change. At the same time leadership in a particular venture requires a complex range of skills including coping with change, delivering change and specialised expertise, which may not be available. Collective identity gives rise to social capital and facilitates the establishment of networks, which may give access to the requisite skills for local development within individuals. Furthermore, different tiers of government play an important role in releasing local potential and this can be seen in cases where the outcome of local people engaged in development initiatives depended on support from a range of institutions. Where this help was not forthcoming local efforts to participate in territorial identity construction were not so successful. While horizontal linkages are important for new groups and networks, the quality and quantity of vertical linkages (including links to central and local government and development agencies) proved crucial for the development of sustained local networks. Both types of networks should be cultivated from the start of a development process.

We could also add that in Sweden, Finland and Scotland, development expertise was often provided through local government (Sweden and Finland) or through local development and enterprise agencies (Scotland) and this contributed to network maintenance and more successful development processes. In Finland local government has shifted its role over the last decades to that of convenor and facilitator of development processes, but this is not the case in all the study areas. We therefore agree with Warner (2001), who points out that in places where horizontal social capital and democratic governance structures exist, governmental interventions will increase the development of horizontal social capital, which further reinforces social capital and democratic political structures. In this environment strong and diverse identities are fostered. Conversely where autonomy of government is strong but linkage with civil society is weak, predation of government by powerful interest groups is more common and community organisations lack autonomy or linkage. Synergy between state, society and democracy is weak. Where civic infrastructure is weak, government can help by decentralising programmes and developing a facilitative participatory structure (Crocker et al. 1998, cited in Warner 2001).

The issue of identity focuses our attention on how a more explicit consciousness of personal and collective identities makes active participation in endogenous development more likely. For example, constructing new identities at different spatial levels may lead to awareness of different types of local resources that can be exploited and marketed in novel ways. A desire for homogeneity and the avoidance of controversy can be seen as a weakness of identity and can lead to conformism, passivity and a lack of collective engagement. Avoiding controversy is also perhaps unrealistic in an era of globalisation, migration, displacement of peoples and increasingly fluid physical boundaries. In all identities there are common elements that can be cultivated to form the basis for cooperation and solidarity. Social embeddedness and participation can encourage diversity and development and do not necessarily lead to uniformity. Social capital and networks can be seen as the outcome of the interaction of collective identities, and yet they cannot be isolated from the whole context that frames local development efforts. The discussion of identities, therefore, cannot be isolated from the situations in which people articulate them.

Chapter 7

Using Environmental Resources: Networks in Food and Landscape

Jo Vergunst, Arnar Árnason, Ronald Macintyre and
Andrea Nightingale[1]

Introduction

This chapter examines some of the ways that European rural environments are
being perceived and used in the context of a general decline in agriculture. In
particular, we explore how networks – our tool for articulating social capital as
an analytical device – serve to link people with places and resources, and with
each other. The links of people to places, or environments, can be seen as an
extension of the concept of social capital that is generally not remarked upon
in the rural development literature.

  Although traditional forms of farming and forestry continue to be
economically and socially important in rural Europe, there are a number of
new and different land use practices are emerging. We believe these novel uses
of environments are important both to future development strategies and in
maintaining crucial links to the past in rural areas. Our main research question
centres on how people in the case study areas (for introductions to which, see
Chapter 1) are using their environments in novel ways and how this is related
to the production and maintenance of social networks. Our topics are diverse,
and include the ways in which new types of food production, renewable energy
projects, sports and leisure pursuits, and nature conservation and management
can contribute to rural development. From this broad base, we discuss the
relationship between the rural environment and the symbolisation of place
and identity in our case study areas. We argue that although promoting locally
specific environmental and cultural features can be a tactic for successful rural
development strategies, it is also necessary to consider some of the effects on
the meaning of society and identity within a place to understand the overall,
long-terms impacts on development.

1   The authors would like to thank all the people in Skye and Lochalsh in Scotland
who gave their time to our research. Our National Advisory Group also provided very
useful feedback. The other member of the Scottish RESTRIM team, Professor Mark
Shucksmith, also worked on the research. The authors gratefully acknowledge the help
of our colleagues in the RESTRIM project around Europe. Some parts of this chapter
draw on Lee [Vergunst] et al. (2005).

Many of the people that the researcher teams worked with articulated strongly the importance of retaining valued aspects of their environments, and, through that, valued aspects of social relationships and their identities. In many cases, this takes the form of trying to retain the landscape as productive land, by linking what might be termed the old and the new in land use. We also argue that rural development will not be successful without carefully negotiating the contradictions that may exist between new and old uses of the land.

Yet in considering these issues, it is crucial to recognise that there is no single 'environment' that has a predetermined set of resources within it. Instead, people perceive resources in their surroundings according to their cultural background, the social and economic networks they take part in, and their own ideas and strategies for making a living – all of which in turn may be influenced by multi-scalar political contexts and economic trends. Thus, following a wide range of social science research (e.g. Castree and Braun 2001; Hirsch and O'Hanlon 1995; Ingold 2000; van Koppen 2000) we emphasise the potential for different groups of people at different times to relate to their environments in different ways. Ingold (2000) puts this in terms of *affordances*, in that the perception of opportunities available in a place depends on the varying habits and practical activities of the user. This is also the rationale for researching what we term as 'environmental' resources, rather than 'natural' or 'cultural' resources. What one person or group understands to be natural in the environment may be seen to be as cultural by someone else, and the ways in which places or objects are described as natural or cultural are in themselves of significance to how development takes place. We prefer to speak of the environment, in the sense of 'that which surrounds us', as a more neutral term. Undertaking research into land use from this perspective allows us to link together many aspects of rural development whilst retaining a concern for actors and the agency of people and communities.

Representing and Practicing Rural Environments

Representations of rurality – the idea of the rural – have been very important to how social change and development take place in rural Europe. It is important to recognise that development issues are not just centred on economic or socio-economic questions, but also include ideas about how ways of life are understood and valued. Later we look at how representations of the environment are used in food production and alternative uses of rural space to reconfigure resources and networks. Now, we will briefly examine some of the historical background to these issues.

The relationship between rurality and nationalism links representations of communities to their environments. Ideas associated with 'the country' and 'the city' have been important structuring concepts for society in Europe throughout its history. The loss of tradition, culture and a way of life in rural areas has been a recurrent theme in literature since the Renaissance – masking, as Raymond Williams noted, the reality of more subtle and contested changes underway in rural areas (Williams 1973).

Schama (1995) and Olwig (2002) point out how 'nature myths' in Europe have taken the form of nationalist stories that link the rural with the nation. For example, in 19th century Britain the idea of a national rural idyll provided a symbolic base for imperial expansion, while in France, the presumed unity and national significance of the 'peasantry' and their rural way of life has supported claims in the present for continued farm subsidies (Hoggart et al. 1995, 18). This conflation of nature and nation also explicit in the English word *country*, which has a dual meaning of nation and rural space. Equating the nation with rural space has significant implications for economic development in those areas. In many respects it is this association that has promoted tourism in rural areas, but it also constrains the kinds of economic activities that are seen as appropriate to what are sometimes thought of as 'pristine' (and also remote and marginal) rural areas. We should understand the importance of rural space and environments not only to people living in those areas but also to an abstracted ideal of the nation held by people throughout Europe.

This suggests that we need to come to terms with the relationships between the different scales of identity that are possible in regard to environments. Many people, and particularly in rural areas, describe some kind of 'link' to the land. They have a sense of who they are that is based on where they are and the practical activities they are involved in. What the scale of that link is, however, may vary according to the type of networks that are most important to them at particular moments in time. People may draw on their identity as produced by their immediate home area in one context and in another, draw on a larger, regional or even national or supra-national identity. Similarly, people may be keen to use their environments to distinguish their local area or region from others within the country, and from other regions in Europe. Development processes have encouraged people in our case study areas to cultivate and use local identities for development ends, and these are often based on images of 'local' environments.

These ideas can be understood as a demand on rural environments which may be contested between different scales and between different networks of people. The emergence and structuring of a 'European' environment, which has its own set of representations, can also to be traced. The relationship between local, national and continental scales of representation and rural development is discussed further later in the chapter, in relation to branding.

Through these links with the environment, we can see how social relationships are always situated in particular places. The literature in social capital usually takes such relationships to exist independently of any grounding in place relations (as part of a wider separation of culture from nature), and we could critique the approaches to social capital outlined in Chapter 1 in the same manner because they are not able incorporate the links with environments that we are concerned with here. Drawing inspiration from actor-network theory (Latour 1986) as well as the idea of affordances, we can look at how particular aspects of the environment are enrolled in the livelihoods and development strategies of rural people. So in this chapter we explore networks, trust and cooperation in such a way that includes the features of the environment that are often so important in development. We aim to take account of how people mobilise both symbolically and materially particular aspects of their environment to build and maintain social relationships and the capacity for collective action. In this way social capital must be placed within not only social but also environmental contexts to understand its role in development.

Linking the Old and the New in Food and Landscape

Many of our informants articulated a strong need to retain valued aspects of their environments, and through that, valued aspects of social relationships and their identities. This often takes the form of trying to link what might be termed the old and the new in land use. Here we discuss a few examples of how people in our case studies attempt to retain these aspects of their landscape through keeping the land productive. We begin by discussing examples of 'quality' and 'local' food production, drawing attention in particular to their role in the reproduction of valued social relationships. We then discuss examples of productive landscapes that are not related to food production. Here we focus on a wind farm being constructed in the Scottish case study area that is seen by some as maintaining the productive aspect of the landscape and economy. Given the decline in agriculture, the wind farm allows the land to be active, and helps to uphold the relationship between crofters and their land. We compare this to the Dalhalla concert venue in Sweden, another example of how an existing resource is being re-used in a novel way to promote development.

Quality, Local Food Production: Networks and Collective Learning

Across the case study areas, and in common with much of rural Europe (e.g. Ray 1998; Tregear 2003), emphasis has in recent years been placed on the production of 'local', 'quality' foods. This emphasis can be seen as a response to the challenges of the changing global agricultural market. Marginal rural areas that are distant from sources of necessary raw material and lucrative markets and where the environment is often harsh find it difficult to compete economically with areas currently more favourably situated. To give one example, while sheep production has in the past been very profitable in Skye and Lochalsh, and may become so again, it generally cannot currently compete with the larger farms on the British mainland, let alone with the huge ranches in New Zealand or Australia. Keeping sheep on Skye thus does not often in itself make a significant contribution to the household or island economy, although it may maintain the skills and ideals of crofting. Producing 'quality', 'local' food is an attempt to engage with the new economy by increasing the value of what is relatively limited production, and aiming it a higher end of the market.

Often, the notions of quality and local are entwined in such a way that at least part of what lends quality to the product is the fact that it is identifiably from a particular place, without having gone through a complex series of processing and retail. Other aspects of quality in the standard and selection of produce often derive from this basis. For example, one company in Skye markets seafood caught around Skye directly to top-end hotels in the rest of Scotland and the UK, developing more value-added markets. Having links between Skye and the markets in the rest of country is very important to the company, but it is not simply in terms of shipping products elsewhere. In this case an awareness amongst suppliers, processors and consumers of the whole chain is seen to be beneficial. The founder of the company enthusiastically explained this as a tactic he used to overcome the distance between producer and market.

> Chefs from the hotels that we supply were invited to Skye [and] they brought their families with them. Whilst the adults were given trips with local fishermen to see the catching side their children were given a day to remember with their own special visits. The adults had a most enjoyable and informative day, so too the fishermen, and by the end both parties through this linkage were better able to understand the demands of the other. (Written correspondence with interviewee, Scottish case study)

Bringing the chefs into contact with the fishermen was a way to overcome a sense of separation between the places of production and consumption – that actually, the fishing villages of Skye are not so far removed from the hotels in their various locations. Furthermore, quality is often seen to be bestowed by the use of traditional production techniques, often less intensive, or using skills learnt over generations and situated locally. In these material and symbolic ways, economic restructuring is mobilised to counter the effects of these very processes of marginalisation. In other words, the marginal area as the location of the nation is made to lay claim to quality.

The Swedish case study provides an example of the close links between rural economic growth and the need to claim the rural as the symbol or location of the nation. Since 2001 a weekend-long event 'Food around Lake Siljan' has taken place in September in the Dalarna region in Sweden. Behind the event are an association for the protection of nature around Lake Siljan, the farmers' association, Siljan Tourism Ltd (of which the local municipalities are the main owners), and shops, hotels and restaurants in the area. The general public is given the opportunity to visit local farms and many restaurants in the area will serve only locally produced food. The idea is not simply to market local food. Rather, the event seeks to highlight the connections between the food people choose to buy and the consequences for the landscape. Indeed, around Leksand and Rättvik in the Swedish case study area, the intention behind local animal husbandry is very much to promote a particular aesthetic image of the landscape at the same time as production of food. Around Lake Siljan, then, the land is seen as a resource more than ever before, and in broader ways. The area is heavily dependent on tourism and the landscape is one of its principal attractions. Landscape simultaneously stands as a symbol for the whole Lake Siljan area, which in turn is often made to represent the whole of Sweden as the 'most Swedish' area of the country.

The aims of new networks around food production are not simply economic, and it is in some of their non-economic functions that they may be most important. They are, for example, often educational in a way that can refer to the identities of local people. The Scottish case study provides another example. In 1994 the Skye and Lochalsh Horticulture Development Association (SLHDA) was formed with support from the Scottish Agricultural College, the LIFE Programme of the EU, the Local Enterprise Company (Skye and Lochalsh Enterprise – SALE) and Highland Council. The group undertook trial plantings on a number of crofts to discover and share skills about how a range of fruit and vegetables are best grown locally.

In doing this links are made to the practice of crofting. Crofting is a form of land tenure unique to the Highlands of Scotland that is based on feudal rental arrangements and the sharing of common grazing lands, and pluriactive labour that can include horticulture. While crofting often takes somewhat different forms in different places, it has an evocative and often painful history and is an important aspect of people's identities (Hunter 1976). One informant discussed how horticultural skills have been developed, or even reinvented, through the SLHDA:

> So the aims were really to replace some of the lost skills, which I as somebody born and brought up in Skye find difficult to identify with because personally I never lost those skills and always carried them on. But other people, young people just didn't do that. ... You have lost about 20 or 30 years, so people of that generation who are maybe in their 30s, don't have the skills, don't have the knowledge, never saw it, or maybe just have a vague idea. (Interview, Scottish case study)

The acquisition of skills can have profound implications for people's identities, allowing them to become legitimate participants in 'communities of practice' (Lave and Wenger 1991) – social relations based on practical activity – into which they were previously not admitted. In this way, a network that was set up to help economic development can have implications for social learning and the way people understand their identity. This contributes to the success of the network in more fundamental ways than the discrete transactions of buying and selling horticultural products.

A further good example of this within SLHDA has been 'Skye Berries', the trading name of a producer in Skye who, although a recent mover to the area, has been able to set up a small business with the help of, amongst others, the

SLDHA. Their 'New Growers Scheme' provided him with a flexible small loan towards his project costs, and also a mentor to help develop his skills:

> Mentoring was a thing that I was really really aware of, that I could plan my thing but as soon as a plant got a disease I would look it up and feel completely unsure as to what it could be .... It was very good having somebody looking at my designs and ideas. ... He thought that what I was suggesting was reasonable and came up with quite useful tweaks to it. (Interview, Scottish case study)

These positive outcomes can be identified as the result of a genuinely endogenous development process of learning and sharing skills. Local control is maintained and relationships are based in socially embedded practices. By retaining control and the acquisition of skills locally, the networks have the potential to provide a link for people to work together again for other causes. Indeed, the SLDHA also successfully run a food van, local box schemes, a weekly market as well as supplying many restaurants with local produce. These activities all contribute to economic development in Skye and Lochalsh not merely through overall turnover but rather through raising the profile of the area and adding value to activities that would otherwise be done as a 'hobby' or in an informal way. Also, these kinds of activities have been a key mechanism through which the skills of people with long histories on Skye and those coming in from elsewhere have been brought together for common benefit.

To summarise, the networks for the production of quality, local foods in these Scottish and Swedish cases are simultaneously educational, social and economic. In Skye and Lochalsh, they can be seen as attempts to maintain crofting – and flexibility has always been part of crofting – and the social relationships associated with it. One key feature of all these relationships and activities is that they seek to continue using the land in a productive manner. But characteristic tensions between 'productive' and 'consumptive' practices of landscape sometimes come to the fore in development projects, and we now turn to this.

Productive Landscapes: Scottish Crofting and Renewable Energy

Crofting is widely understood by those who take part in it to be characterised by a spirit of cooperation and sharing, although, in something of a paradox, it is often evoked in the context of these characteristics now being lost. But this perception of decline allows new initiatives to fall back on this imagery in attempts to create new networks to promote development. This situation lies behind the following comments, discussing a recent initiative to produce and market local food in Skye and Lochalsh:

> Maybe 50 years ago when people were crofting and they worked, they had much more of that informal network, in a lot of ways it is really just, you know, reinventing

crofting. It is not anything new, it is just going back to what worked in a rural area. (Interview, Scottish case study)

'Crofting' in this sense is a set of social relationships in which people worked together, often informally. To 'reinvent' crofting is to produce the quality of these cooperative and communal working relationships. Clearly, crofting has to an extent become romanticised, and images of crofting in the past are often very much of a rural idyll in which people worked in harmony with each other and the land. Elements of this discourse can be found in many of the interviews carried out in the Scottish study area, in which, often, older or retired crofters describe the lack of close relationships now, in comparison with what they had known when they were younger. The decline of crofting can thus serve as a convenient metaphor for some of the other social changes that have occurred, such as the influx of people new to the area. New food networks are thus not only a reaction to but a commentary on these changes.

This became especially pronounced in debates that arose around the construction of a medium-sized wind farm in a crofting township on Skye. Permission was initially granted by Highland Council for 27 100-metre high wind turbines to be built, but the decision was contested, and only in 2007 was a smaller proposal accepted. The wind farm is to be established, owned and run by a multinational engineering company but the project was the result of the company's cooperation with the estate owner and many of the crofters on the estate.

The wind farm proposals have been very controversial. The benefits projected to flow from the wind farm to the crofters and the township are mainly economic. The company building and running the wind farm will pay rent on the common grazing land where the wind turbines will be located, a proportion of which will go to individual crofters (mandated by crofting regulations). In addition, a community fund was to be established that, at the time of our research, was to generate at least £25,000 a year for the local community for 25 years, although the terms of these payments were still being negotiated.

Meanwhile, objectors to the wind farm proposals argue that the development will compromise the beauty of the landscape and consequently harm tourism, which is vital to Skye as an area. The Skye Wind Farm Action Group, the mouthpiece of opponents to the wind farm project, and individual objectors emphasised the pure, untouched, natural landscape of Skye. The group stated in one of its publications:

All the policy and advice notes from government level down acknowledge that wind farms do have a radical effect on the landscape, particularly areas of a largely natural and static character bearing little evidence of man's activity and impact as are involved on Skye. (http://www.sw-ag.org/Gallery/Letter%200.htm)

The Group argues that the wind turbines' height, pylon diameter and sweep area would make them very 'intrusive and impossible to ignore in an otherwise static landscape subject only to natural movement.' Because of this the wind farm would inescapably compromise the 'unspoilt character of the landscape and environment'.

Figure 7.1   View of the site of the proposed wind farm in Skye, Scotland

The views of many of the crofters living in the area were very different. To them the land is and should be kept productive, and the attraction of the wind farm is precisely that it makes it possible to keep the common grazing land productive. It becomes almost a way of resisting changes in the global economy that are affecting the viability of agriculture and thus the productivity of the land. One informant summed this up by saying that Skye 'is a working island and everybody ... here has got to make a living, they have got to survive'.

The importance of this lies partly in the strong tie between the land, their identity and their community that crofters in this area articulate. They base this claim on an understanding of the landscape as anything but static. To crofters the land is historical and changing, it is fundamentally created through the activities of human labour through the generations. One interviewee put it in the following way:

Much of what you see has been created by the activities of people over hundreds and hundreds of years. We have lived here all our lives. There are these ties between the land and the people, between what we would call our crofting land and the people. ...)There is a tie between us and the land.

This tie is so strong partly because it is also a tie between people, mediated through collective work in the landscape in sheep and cattle farming, forestry and horticulture, for example. Crofting as a set of activities links people to their predecessors and contributes to their idea of who they are. Indeed, it is a tie that refuses to distinguish 'us' and 'predecessors': 'we have lived here all our lives'; 'all this land about me here is ours, it's mine', our interviewee said. The perceived need to cooperate and share is articulated by crofters as important to crofting as a way of living and a valued feature of their community. Asked about the notional consequences of the disappearance of crofting, one crofter said:

> It would be a catastrophe for crofting communities. A lot of the community spirit would go because at the moment the backbone of the community is the crofting and the fact that there are active crofters here. That is basically what keeps things going. We have meetings and go out together, we meet regularly which if it wasn't for crofting you would probably only see your neighbours every so often. But it brings all the working crofters together on a regular basis; it has got that community spirit, which is good for the community.

Landscape, to local crofters, is thus not just plants, soil and animals, and much less 'scenery'. It relates to work, heritage, social relationships and land tenure, and it is thus a political as well as personal and communal entity. It is these ties that people draw upon to promote economic development and the cooperation necessary to begin new enterprises. Without the sense of community that is based in these practical activities, many of the features that hold people in Skye and Lochalsh, as well as attracting tourists, would be diminished.

Maintaining a productive landscape is seen as very important for crofters in the our study area. Their active engagement with the land allows for the reproduction of valued aspects of social relationships of cooperation and sharing and because of this, the active, productive landscape is emphasised: the environment as created through human labour. The opponents of the wind farm, in contrast, articulate a sense of the land as 'untouched by human activity', unspoilt and natural. This is used as a basis for opposing the development of a wind farm, while allowing for the consumption of the landscape as scenery through lifestyle choice and tourism. We therefore have two different views regarding the very meaning of land and landscape which affect the kind of rural development that is thought to be appropriate. We can turn again to the Swedish case for a comparative example.

A Swedish Rural Concert Venue

As described in Chapter 2, in the early 1990s a former opera singer was looking to build a new opera venue somewhere in Sweden. Having a summer house in Dalarna she happened upon a disused lime stone quarry and became convinced that this was the place she had been looking for. Using her extensive network with leading people in Sweden's cultural life and mobilising local networks of volunteers the opera singer managed to create the concert venue, Dalhalla, which is now one of the main tourist attractions in Dalarna.

One of the most interesting aspects of Dalhalla is the different set of people it brings together. On the one hand the running of the opera is made possible by extensive networks of local volunteers whose considerable work is rewarded by benefits to their local community, and whose contribution is perhaps made more possible by existing traditions of voluntary engagement. On the other hand are people from outside the area, and mainly from Stockholm, who frequent the opera productions and some of whom have paid handsomely to have permanent seat reservations and their names put on a chair in the arena.

We can make a further distinction in the ways in which different people relate differently to Dalhalla as a landscape. In an interview a local volunteer explained his involvement in the running of the opera:

> Yes, Dalhalla ... people have been working there in the limestone factory, who made the pit, so to say. It is a natural connection. (Interview, Swedish case study)

In this narrative Dalhalla was made by the people who worked in the limestone quarry and this interviewee sees volunteering there as a natural continuation of that work. Others emphasised the sense of a collective identity in Dalhalla that is not so different from that of the Scottish crofters, in forming the landscape through collective labour.

This understanding is in quite sharp contrast to the way in which the formation of Dalhalla is presented to outsiders by the company that runs it. Here emphasis is placed on a meteorite that fell on the area millions of years ago and contributed to the formation of the rock strata that was later quarried. The way visitors are told the story now sounds almost as if the quarry itself was formed by this crash. The creation of the quarry that has become a concert arena is presented as a natural process, and the human labour and social relations that created the setting are not to the fore. The setting is made 'natural' and presented as a stage for the culture.

As an aside, the two different names that the limestone quarry has been known are also interesting. The original local name given to it was 'Draggängarna'. 'Dragga' means to drag something from the bottom of lake, for example, while 'ängarna' means meadows, indicating a damp ground perhaps from which something is dragged. 'Dalhalla' meanwhile is composed of 'dal' meaning valley, and 'halla' meaning palace, evoking the presence of royal splendour and power in the midst of the most Swedish of Swedish countryside. The landscape and nature of Dalarna are often presented as a symbol of Sweden, and are the setting for the production of a very international cultural form: opera. These symbols thus position people and landscape in different ways. The international, 'naturally' formed image of the concert venue gives the place a 'pure' nature aspect, whereas the limestone quarry image is one in which people worked collectively, often in difficult and even exploitative relations.

These conflicting images can be placed against the voices of local volunteers, who in interviews for this research sometimes talked about having problems with the amount of work they feel required to contribute to Dalhalla. They speak of 'involuntary voluntary work'. There have been some attempts to mitigate the sense of unfair compulsion by providing funds to local village associations in return for voluntary work. Meanwhile, the people who maintain permanent reservations and have their names on chairs in the arena are mainly urbanites for whom the venue is an escape and a place for finding their Swedishness. Portrayals of the landscape serve to present or hide particular forms of relationships between people and their environments, and between different social groups, and these can certainly play into rural development processes.

These examples illustrate the value of a local historical perspective on rural development. It is important to understand that how the local environment is understood is part of the social and economic characteristics of an area. In a beneficial sense, joining together traditional and new practices in regard to the environment can maintain those aspects that are locally valued and valuable, whilst allowing for change and development. The maintainance of crofting skills in Skye and Lochalsh is a positive example of this, while the tensions over the wind farm, and voluntary work around Dalhalla, show up some of the complicating factors.

Linking the Old and New

The notion of a productive landscape is a good summary of how many people in our case study areas idealise their relationships with their environments. Although many people have a strong desire to maintain agricultural production, there are ways in which a productive landscape can be formed that do not rely on growing heavily subsidised, mass-commodity foods. Integrating local and extra-local quality food networks is a good basis for this, in which collective learning will be an important factor. The two examples of a wind farm and a concert venue may seem far removed from traditional development practice, but for the local people involved they maintain the functions of upholding social relationships, a sense of identity, and a strong local economy that are vital to society anywhere, rural or urban. Activities that are simply 'service' based, such as some forms of tourism, for example, may easily be seen as undermining the close relationships with local environments that are important to many people in our cases studies.

At the same time there are a variety of ongoing conflicts as these new initiatives become more widespread. So far, we have described the tensions surrounding the negotiation of a Scottish wind farm, where perceptions of the local environment, local rural development, national public energy policy and international capital are in contact with each other and will inevitably pull in different directions. Conflicting understandings of what nature and traditional culture are, are played out in localities. The idea of 'involuntary voluntary work' amongst the residents around the Dalhalla concert venue also raises questions of who actually benefits from these projects. The desire of rural areas to be attractive to those in cities is understandable, but if there is to be a 'transaction' between them, it needs to be on fair terms. In the next section, we extend our look at the relationship between representation and practical activity in rural development in examining some networks for the branding of our case study areas.

Images of the Environment in Marketing for Rural Development

In this section we continue thinking about how environments in our case study areas are imagined, this time as part of broader marketing strategies that are becoming common in rural Europe. Our aims are to look at what the areas are made to stand for and who has control over the process.

Producing Meaning in Marketing: Intentions and Side Effects

In all of the regions we have studied, images of the environment and culture play an important role in how the areas are presented to potential tourists. Such representations can also affect the ways in which social identities are constructed, and even how the areas are mobilised as symbols for economic and political purposes. It is important to understand these issues if the potential negative side effects are to be avoided. In the academic literature, attention has been paid to how identity and culture can be instrumentally used in these ways to secure development (e.g. Ray 1998), but we contend that a critical perspective is necessary that evaluates the relationships between culture and development. The two issues that we consider here are the content of such representations and, in the subsequent section, the social networks that produce them.

The Irish study area is notable because of the efforts there to construct a new region and a new regional identity around a particular feature of the environment: the lakes. The purpose of the exercise has been to enhance rural development and local businesses. The creation of the Irish Lake District deliberately evokes the much better known English Lake District in order to attract tourists. Through its construction around a particular local feature of the environment, the Irish Lake District only stands for and represents itself, rather than symbolising the nation. However, the rural west of Ireland generally is often seen to be the cultural heartland of Ireland, and the Lake District attempts to appropriate some of the images of the generalised 'west of Ireland' for itself.

Similar processes have been in operation in the Norwegian study area where another feature of the environment, the mountains, has been mobilised in a rural development strategy. Chapter 5 describes how the Mountain Region Council was formed to act as a forum for cooperation between smaller municipalities in the area, and efforts were made to promote a sense of identity towards the region. Again, the main emphasis here is on another feature of the environment, rather than local culture or any specific traditions, as the distinguishing mark. The Norwegian Mountain Region is not mobilised to stand for Norway or the Norwegian nation. Rather, like the Irish Lake District, it stands for itself as a way to promote the area by having a distinctive identity different from other areas of Norway.

As in Norway and Ireland, in the Finnish study area of Sotkamo, which includes the village of Vuokatti and a regional sports centre of the same name, the symbolic emphasis is on the natural environment, the hills and the woods in particular. It is stressed that the area is a harsh environment and its inhabitants are moulded by its hardships. The area is simultaneously renowned for its beauty and has frequently been evoked in the creations on many of Finland's most celebrated artists. Through these symbolic linkages the area is sometimes made to stand for and represent Finland as a whole and it is in many ways appropriate that the Finnish national ski team should train in the Vuokatti area. This image has undoubtedly benefited the area as a tourist attraction. Recently attempts have been made to create a more streamlined and homogenous image, stressing tourism, recreation and the sport facilities of the Vuokatti Hill area with its hills and its lakes.

Turning the image of Vuokatti Hill into a brand has utilised the earlier images and symbols. The old logo of the local sports club contained an image of Vuokatti Hills. The new image has a stylised wave profile of the hills. The municipality of Sotkamo has adopted this new image as its logo as have the biggest tourism enterprises in the area and the Sport College in Vuokatti. The Finnish case is an example of a successful attempt to create a powerful and locally and nationally accepted brand.

The Isle of Skye, in the Scottish study area, has for a long time had a very powerful image. It evokes the crofting tradition of the area and the island's place in the often dramatic and violent unfolding of Scottish history. In many ways this image also rests on the natural beauty of the island, the purity of the air and the water and the harshness of the environment. The dramatic mountain range of the Cuillin Hills, in particular is central to the vision of natural beauty on Skye and large parts are formally designated by the Scottish nature conservation agency as 'Areas of Outstanding Natural Beauty'. While Skye is not exclusively made to stand for the Scotland as a whole it is an important part of a wider representation of Scotland and Scottishness. Skye is thus represented as being more than its immediate self.

This strong image is seen by development professionals as important for the economy of Skye. It attracts tourists and it has been utilised in marketing products from the area. Efforts have been under way to make even better use of this and brand Skye. To these ends the Skye and Lochalsh Marketing Group (SLMG) was instigated by a small group of businesses in the area and received funding from the Local Enterprise Company. It began with a series of meetings and local publicity that encouraged involvement from businesses, public sector bodies and community organisations. When the scheme was formally launched in early 2003 a distinctive logo was unveiled, featuring an image of a pebble and the words 'Skye the Island and Lochalsh'. Businesses were able register to use the brand image for £25 and put the Skye and Lochalsh label on their products. SLMG hoped that this promotion of the local products and services would help increase the revenue generated in the area itself.

In its efforts the SLMG has stressed the natural environment of Skye and Lochalsh. In one leaflet it describes the brand, as: 'Timelessly appealing, universally understood and formed from a beautifully simple concept developed over a mere 50 million years' – tying the brand to an idea of the landscape of Skye. An image of sunset over the Cuillins, where only their outline was visible, was used in publicity to launch the brand. The Group's logo, a pebble with an outline of the Cuillins drawn in white Skye marble, similarly emphasises the natural environment. Interestingly, the image bears a striking resemblance to the stylisation of the Vuokatti hills in Finland (Figure 2.4) and also the logo of the Irish Lake District, which features a curved green line over a flat blue one.

In all these cases, features of the environment are used by marketing organisations because they appear to be socially neutral. Mountains are used in local branding in the Scottish, Norwegian and Finnish cases, and lakes and hills in the Irish and the Swedish case as well (although the later was not articulated so explictly). In our Italian case of the Maremma region the image chosen was a wild boar. Such images can be used as symbols that seemingly everyone in the locality can have equal access to and ownership of. This is in contrast to the difficult decisions that would be needed for a 'cultural' image. One of the organisers of the Skye and Lochalsh Marketing Group conveyed this argument when describing the difficulties they would have had in using an image with a more cultural basis, such as a person:

> It has to be something that can be fairly simple and it can be transferred from different things back and forth. So if you had taken a person, be it a male or female, adult, you know, old? Skye born and bred, or do you get a sort of nice model? Too many problems. (Interview, Scottish case study)

While no doubt beneficial to local businesses, such branding can also be problematised. The image of Skye being constructed through the branding is of an untouched, unspoilt, static natural environment. The effects of human living and labour can be hidden, or even ignored. Our discussion of the wind farm proposals in illustrated the contrast between this and way in which crofters construct the landscape and their relationship to the land. For them, an active landscape is essential for the reproduction of valued social relationships, social structures and identities, and in maintaining the landscape as it now is. Many crofters emphasised to us how it is their agricultural activities that have produced the beauty of the landscape.

The ways in which these features of the environments are being used in images is becoming an important part of development strategies in rural Europe. There are other social relationships that can be hidden behind the images. It is not just 'natural' features that are used in branding, however, and in the Swedish and Italian cases, in particular, 'cultural' aspects of the locality are also used for branding purposes, such as the Dala-horse in Dalarna in Sweden, despite some the problems outlined by the interviewee above. Indeed, as the crofters themselves have argued, the 'cultural' features are integrally a part of the 'natural' landscape. The kinds of images that are used, and how they circumscribe a particular set of development options, have significant implications for rural development strategies based on branding.

In the Swedish study area of Dalarna, the landscape, but more specifically the cultural traditions associated with the area are represented as the most Swedish of Sweden, and in this way made to stand for the country as a whole. This image again is clearly important for the local economy in attracting tourists and newcomers to the area. But others are more ambiguous towards these commodifications of landscape and culture. It could be argued that through these symbolic constructions their ongoing lived reality is lost, and they become features in a kind of national theme park that enacts Swedishness for visitors.

The Italian study area in southern Tuscany is often presented as being marked by steep hills and deep forests (Cecchi and Micocci 2002). These hills and the forests are, in turn, made to appear as one of the most important sources of the region's well-known cuisine. Stressing that traditional methods are employed to convert raw materials into food, a 'natural' environment is thus made to appear as a wellspring of a cultural tradition.

The pre-Roman (Etruscan) past of the area is also emphasised through this. Many of the monuments from this period are open to the public and the use of them in tourism marketing also contributes to a sense of identity in the environment. In this case, the environment is understood as cultural and historical, where the past and present are conjoined in ways that are not present in areas where a more 'nature' based strategy for branding is used. The past becomes part of the representation of the area but against a backdrop of larger story of Roman cosmopolitan progress. In this way, it could be argued, the area comes to represent a chapter in the nation's history. This shows again how the choice of cultural markers has implications for how the identity and development of an area is understood.

The Social Networks of Branding

Having described how the areas are variously represented as sites of nature, culture or nationality, or a combination of all three, we conclude this section by examining briefly who it is that can represent the areas in question. The social networks that are involved in branding are significant for the meanings constructed through it, as well as for the control over local development processes.

At least two of our areas, in Norway and Ireland, did not have a particularly strong image or identity on the regional scale until recent development efforts. The attempts to create such images were instigated without any formal pressure from the outside. It appears to be working in the Norwegian case because of strong support from the municipality, helped by the fact that in Norway municipalities are quite strong entities. The Irish initiative, meanwhile, has been the responsibility of a local enterprise company and has had problems securing continued funding and support.

The opposite of this would be the case for Sweden. Dalarna has a very powerful image, one that is largely constructed outside the area, and one that is made to stand for the rest of Sweden. While helpful in attracting tourists and newcomers to the area, the image can be interpreted as disempowering to local people and possibly even limiting their choices of development routes. The same is in many ways the case for the Finnish area and for Skye, both of which have for some time possessed a very strong and well-known image. This is a image in many ways largely beyond the control of local people, and because of the stress put on nature and natural beauty in the image it can be seen as being quite different from the ways in which local people construct their relationship with the land. The Skye and Lochalsh Marketing Group can be seen as an attempt to wrestle control over the imagery from outside forces. Similar processes were at work in the Finnish case of *Aito Maito* (Authentic Milk), as described by Lehto and Oksa in Chapter 2. In both cases such branding or marketing has been largely locally-driven, but in Skye and in the Irish Lake District there has been debate over how far the relatively small network of people driving the process can claim to stand for the whole area. In the Skye case, the branding has tended to stress untouched natural beauty in its objectification of the area, portraying an image of the region at odds with that often articulated by local crofters.

Representations of local areas, such as those found in regional branding, are important elements in contemporary development processes. They are elements that can, no doubt, considerably enhance an area's attractiveness as a tourist destination and the lure of its produce. Representations and brands do not however arise by themselves, of course; rather they are created by networks of people. While these networks may strive towards inclusivity, representations are always, unavoidably particular, as the Finnish and the Scottish cases suggest strongly. The content of an area's representations or brand is similarly important, because it becomes part of the opening up and the closing down of further development choices. Similarly, creating an image of an area as somehow representing the nation may curtail the potential for self-determination that local inhabitants have in their development strategies. It is, in summary, the diversity as much as the similarities of marketing strategies for development that is most striking amongst the rural areas that we have studied.

Networks around Nature Conservation

The focus up to now has been on how the environment gets explicitly caught up and used in rural development projects. But in many of our case studies we came across examples of nature conservation activities that had a bearing on the course of local development, and which often related strongly to our themes of continuity and change in rural areas.

Nature conservation does not of course happen 'naturally', nor in a reflex response to external influence. Rather, it is planned, carried out, and as we shall see, contested according to local circumstances. In the case study areas it is apparent that the scales on which nature conservation takes place vary widely, and, as a result, has varying impacts on rural development. Patterns of land ownership in the different countries also affect both nature conservation and the overall opportunities for new uses of environmental resources in rural development. While previously we discussed the continuing importance of food production and how, for example, wind turbines can be seen as relating similarly 'productive' sense of landscape, here we consider a set of activities based in the environment. This engages again with some of the processes that may be marking a 'post-productivist transition' in the countryside (Ilbery 1998), and, specifically, with the possibility that consumptive (such as tourism) rather than productive land uses are becoming more significant in the use of rural environments.

The central question at this stage is how far the networks that surround these novel types of land use are important to rural development. Academic research on rural environments has usually focused on 'threats' to the environment, in terms of pollution and habitat loss. We are not, however, primarily investigating the kinds of environmental degradation that are underway in Europe, although environmental concerns of course need to be central in all development, which is usually now understood in terms of models of environmental sustainability. As Hoggart et al. (1995, 231) write, most important in regard to European rural environments have been degradation linked to agriculture, the extension of urban activities into rural areas, and the ecological and landscape implications of tourism. We might now add to these increasing concern over climate change, including sea level rises and desertification. In order to ameliorate some of these problems, there have been a wide range of efforts to conserve rural environments, and in many cases these also contribute to rural development. It is the overlap between conservationist and development processes that is of interest to us here, and we might expect the tensions between the two to be significant for how rural development takes place.

Conserving Local Natures

Although all the case study areas in this project are covered by legislation to protect the environment, we found a considerable range of nature conservation and landscape management initiatives that also reflected local priorities and circumstances. It is clear that the social networks that are formed around nature conservation do vary considerably, and yet 'the environment' emerges in many cases as a strong rallying point for collective action. The extent to which the activities of nature conservation are seen to be encompassing the interests of the local, as well as national or international criteria for environmental protection, is often key to whether it will be taken on enthusiastically by local people, and then potentially made part of their local development activities. In this section, we will examine some of these different scales of nature conservation and landscape management activities.

In the Swedish case study areas of Leksand and Rättvik in Dalarna, local people became concerned when they realised that the decline in numbers of livestock around Lake Siljan might affect their views of the lake. If there are not enough animals grazing in the fields, then shrubs and trees begin to grow up, blocking the views that are valued by residents. Maintaining both the tourist economy and the property prices in the area lends a practical concern to 'keeping the landscape open', as it is known. One solution has been to organise grazing associations. These are voluntary groups that work to clear the ground and build fences, and farmers are then invited to keep cattle there in order to keep the shrubs down. One group has even bought its own cattle for this purpose. The grazing associations were started by people who were not linked to the agricultural sector, but were nonetheless concerned with the environment. Some municipalities are also taking on the work of clearing trees, using both machines and cattle to change the land cover. In one case, EU Objective 3 funding has been used to provide a training scheme in this work for long term unemployed people. That the agricultural sector can be seen as a solution to environmental and landscape problems, and including a direct social benefit as well, shows how these sectors are by no means necessarily antagonistic. In this case, the strong local voluntary networks in conjunction with active, local-level municipal authorities were able to facilitate a local dimension to landscape management.

In the Scottish case study area, trees have also become an environmental and rural development issue as crofters are able to apply for money from the Department of Agriculture for planting trees in their Woodland Grant Scheme. Native species such as willow, ash and birch are relatively rare in what is already (in contrast to the Swedish case) a very open landscape, and they have the advantages of providing shelter to livestock and habitat for many bird species. As with the Swedish case, the scheme has encouraged a very local level concern with improving the environment in ways that are valued by a number of different groups of people. These include crofters, other local residents, tourists, and nature conservation organisations and their members. The character of the networks surrounding a particular crofting community in regard to the Woodland Grant Scheme was discussed by one informant in the following way:

> In communities you do need individual people that will be leaders. So it's just made up of a lot of different individuals and different interests but the important part is that they are able to work together and as I said I think at the beginning, don't get me wrong that everything just goes hunky dory and that there is never any dispute over anything. There are some big disputes at times. But there will be a big dispute and that's it.

These comments show how the community was able to work together, and thereby deal successfully with these various organisations as well as the decisions over how extensive the trees should be, where exactly they should be located and so on. Dispute resolution – or at least not bearing grudges at the end of disputes – is highlighted as an important factor, and this might be particularly important in nature conservation, which has been the cause of much controversy in rural communities in Scotland and elsewhere (Lee [Vergunst] 2007). In this case the land that the community lives on is owned by a nature conservation organisation, the John Muir Trust (JMT), rather than a private landlord. The crofters were able to start their forestry work with help from the JMT, whose interest-free loan to the crofters allowed them to build fences and thereby access further funding for tree planting from the Department of Agriculture.

Comparing these Swedish and Scottish cases shows the importance of a real local engagement with environmental schemes. Although the latter was an example of joining a pre-existing scheme, in contrast to the former which was formed from the ideas of local people, they both required significant local cohesion and control. However, this 'social capital' was not a free-floating or abstract phenomenon, but was formed through the specific interactions between people and their environments. There was a desire to produce a particular kind of landscape, and community animation followed from that.

Furthermore, the interactions people had with other organisations were extremely important. Environmental or landscape improvements usually have no direct market value, unlike food production, for example, and yet often play a central role in rural development and maintaining quality of life in an area (itself critical for rural development). Alternative or non-market-oriented means of funding them therefore need to be found, and these two cases show the considerable efforts that need to be undertaken to achieve this. This is despite the uncontentious nature of the projects, at least in their principles. The problem here might be in translating the support on virtually all levels of society (local to international) for improving environments into specific and manageable mechanisms that communities can engage in, because they may be unlikely to be funded from the more common economically-driven enterprise grants. This comparison also shows the importance of understanding *local* environments: the enthusiasm for tree planting in one area is matched by the enthusiasm for tree clearing in another, and yet both are valid and valuable environmental improvements.

Conflict between Local, National and International Nature Conservation

Disputes can arise between different scales of networking in regard to nature conservation. The legacy of one such conflict is still present in the Irish case study area, when a law was passed to introduce permits to fish in lakes in Ireland. Previously, fishing had been free and open to all on the larger lakes, which were not privately owned by anyone. The law was seen as inducing government ownership of what had been an open and common resource. One interviewee in Ireland described the response of local people:

> Fishermen eventually decided on a peaceful protest. They removed their boats from the lakes for two years and anyone that went out on the lake was ostracized. It was back to the old boycott that was used in the same district many years before. It was the same in a new and practical form. The tourist industry was hit and a lot of money was lost. The incident created a network of strength among people in the region that was colossal. It wasn't 100 per cent but among a community of fishermen and all the people involved e.g. those in the B&B industry it was huge. That strength couldn't have been generated if it wasn't already there in latent form. Eventually the law was rescinded. (Interview, Irish case study)

Local control of the lake was not only important for the management of the environmental resource, but also for the sense of local identity and community that could be produced through it. Attempting to exercise power over who could or could not use the lake was seen as undermining the ability of the community to take care of its own resources, and, ultimately, to undermine the community itself. Through their protest against the law, people in the area were able to effectively mobilise a community strength or spirit that this interviewee understood to be present in the area already. In this case, although it is in opposition to 'outside interference' to local affairs, the dispute nonetheless brought to the fore local networks that are based around an environmental resource.

These are examples of the type of rural development debates that can take place in relation to the local environment. Aspects of the environment often seem to provide significant motivation for local people to get involved in rural development processes, and networks are often formed that include the environment within development strategies.

The debates surrounding the wind farm on Skye in Scotland are also relevant here. While there was been significant opposition to the development of a wind farm in the area, it is notable that the 'environmentalist' lobby in the UK and elsewhere are by no means uniformly against this specific, or indeed other, such proposals. They are generally in favour of renewable energy as alternatives to fossil fuels or nuclear power. However, others argue that the damage to the landscape done by wind farms is not worth the energy that will be produced from it, and different sources of renewable energy should be developed instead. The networks that have formed around the wind farm do, therefore, cross-cut local and national contexts, and the scales that the environment is understood at similarly varies. For some, the local concern for a 'productive landscape' is greatest, and for others the principle of 'damage' to a wild part of Scotland is the motivating factor. Still others locate the wind farm on a scale in between these two, as it is seen to be a threat to the tourism upon which the local economy substantially depends. On a local level, despite the fact that debate is ongoing, the networking and negotiating skills engendered by the situation thus far may stand the community in good stead for the future. We could thus identify a general 'capacity-building' benefit (Shucksmith 2000) from the involvement in the wind farm debates, despite their apparent divisiveness.

The kinds of networks that can be formed around nature conservation and rural development emphasise the importance of particular aspects of the environment for motivating community animation. Of concern for many rural people is the extent to which they are able to maintain control over the environments that are important to them. This can be understood in a variety of economic and cultural senses that should not necessarily be distanced from each other. Such factors tend to reinforce rather than oppose an overall conception of 'an environment'. The tendency of nature conservation discourses, on the other hand, is to separate environments into 'natural' and 'cultural' components, as a precursor to exercising control over them. But we have presented a number of examples here that show how nature conservation schemes and organisations can work effectively with local communities and their economic and rural development strategies. Local involvement and initiative have been important factors in these cases.

Some Effects of Land Ownership on Networks in Nature Conservation

Finally, we would like to discuss some of the underlying factors that are important for how nature and nature conservation are understood and played out in regard to rural development. The pattern of land ownership and access is very different between the case study areas, and it is important in determining the range of debates that are likely to be held in the different places. The ways that land ownership and access affect the forms of social capital and networking that are possible in relation to environmental resources are significant. This is relevant for understanding networks and social capital in that social capital involves 'features of social organisation' (Putnam 1993, 167), and yet it is the links between social capital and particular places that we are exploring here.

In the Irish Lake District fishing conflict described above, reference was made by an interviewee to the time of the landlords, who used to own the lakes and the substantial resources associated with them. Maintaining free access to the lakes was linked to the freedom from landlord-tenant relationships, and more generally to the self-determination of local people. This situation is paralleled to some extent in the Nordic countries involved in our research, where the customary right of access to private land such as forests and uncultivated areas is protected as *allemansrätten*,[2] or 'everyone's rights'. Although the specific activities that can be undertaken as part of *allemansrätten* are limited, mainly involving access (including overnight camping and the picking of berries and mushrooms), the principle of open enjoyment of these rural resources is strong (Colby 1988). It is an embedded tradition that is rarely alluded to explicitly during interviews or other formal research situations, but nonetheless seems to inform and organise attitudes to environments and the perception of resources within them.

On a general level, we can suggest that the presence of *allemansrätten* would be likely to lead to the notion of a common identification with land, which might be tied to the strong tendency towards environmentalism within the Nordic countries (Wiklund 1995; Witoszek 1998). In localised cases, however, the idea of what can constitute a resource in rural development may also be affected by such a concept. One interviewee in the Mountain Region of Norway described how she was not pursuing any plans for making money from her 'summer-dairy' in the upland part of her farm, despite the potential for income generation from sports and leisure activities. Although the buildings were rented out for money, the land itself was open for access by all (for ice fishing in the winter and horse riding in the summer). The ownership of the land was in the hands of a foundation, although the interviewee was in a position to charge rent to the anglers and riding company if she wished. In the latter case, the close friendship network between the riding company and herself precluded any further financial dealings between them. Underlying the situation, however, the *allemansrätten* meant that any attempt to limit access or profit from controlling access would have been seen as inappropriate.

While this is clearly just one case, it does draw attention to some of the contradictions of using environmental resources in contemporary rural development. There is a danger in the efforts to commodify and privatise access to such resources that traditionally embedded forms of social organisation may

---

2   The Swedish term – similar words exist in the other Scandinavian languages, and the equivalent in Finnish is 'joamiehen oikeudet'.

be undermined. *Allemansrätten* is a powerful concept in the Nordic countries, and many would claim it to be part of their cultural heritage even though it is usually implicit rather than explicitly mobilised. If rural development strategies are geared towards the commodification of environments (and their representations, as often happens in marketing strategies), we are presented with a new development problem of maintaining valued ways of life and valued landscapes. The difficulty is with translating the desire for access to such areas and their resources into a tangible benefit to local people in such a way that it does not destroy the very basis of that desire – that the land is perceived as natural and in common ownership. There are parallels between this scenario and that described more narrowly for nature conservation.

The situation in Skye and Lochalsh in Scotland is also interesting in this regard. Many crofters do see themselves as owning the land in symbolic if not usually literal terms. Particularly in the Highlands of Scotland, ownership of land is concentrated in the hands of very few people, who then rent it to occupiers (Wightman 1997; Callendar 1998). However, through the legal constitution of common grazing committees, and some new joint-ownership initiatives between environmentalist organisations such as the John Muir Trust and local communities, a sense of local common ownership emerges. The Land Reform (Scotland) Act of 2003 also gives crofting communities the right to purchase their land if they can come together as as group (which is not necessarily straightforward) (Sellar 2006; Mackenzie 2006). This is underlain by the feeling that the labour expended on the land over generations of agricultural activity gives a form of moral authority over what should happen on it.

The case of the wind farm in Skye shows how recent rural economic changes can be affected by such processes. While the crofters are able to use the opportunity of a new development in the area to assert their claims to control over the land, they do so in the face of the continuing advantageous situation of the landowner himself (who can claim 50 per cent of any rent due on the land), and also in the face of a powerful multinational energy and engineering company. In Skye and Lochalsh, therefore, the opportunities for local self-determination of the use of environmental resources may be increasingly fragile. On the other hand, the efforts of local and regional organisations such as the Crofters Foundation and the local government to secure some financial benefits for the crofters collectively may be an example of how the problem of 'translating' resources from an environment into rural development can potentially be worked out.

Conclusions

This chapter has examined some ways in which social networks produce particular kinds of environments in rural Europe. 'The environment' is not a pre-existing context for development, but rather is constituted through particular social processes. The potential uses and resources that are perceived in a particular environment will depend on how social processes and relationships work. Some of the implications of this for how rural development takes place have been touched on throughout the report, but in this conclusion we draw them together and reiterate what we see as the main issues that emerge.

Specifically, we can note the continuing importance of food production to rural development strategies in many of the study areas. Considerable academic effort has been put into understanding the contours of the 'post-productivist transition' (Ilbery 1998), in which the post-war attempts to continually increase agricultural yields is supposedly being replaced by a more fragmented set of demands on rural areas. The research has shown that the interactions between various productive and consumption-oriented activities are complex and do not support a simple transitional model from 'production' to 'post-production'. Changes to European agriculture and rural development are emerging only gradually and partially, and where apparent, they raise new questions about rural society and development.

We can suggest that in our case study areas there is an unwillingness to substitute an agricultural peripherality, in which they have been subject to the policies and economic forces of mass food production, with a consumption or tourist-related peripherality, in which certain kinds of landscape are produced for the enjoyment of others. Many of the food enterprises and supporting marketing networks in the study areas are based around a territorial approach to rural development in which not only features of the locality are used in marketing, but the efforts are intended to produce common benefits to people in the locality. This is in contrast to what can be seen as individualised relationships between the producer and the state and market that characterise EU agricultural subsidies and commodity production (Friedman 1979).

In maintaining food production in these remote rural areas, there is an important sense that the landscape itself needs to be kept productive. Such a landscape is valued by many rural residents and would result in a significant decrease in quality of life if it were seen to be lost. In Skye and Lochalsh, the efforts of crofters to keep the landscape productive structure their responses to other rural development possibilities. There, crofting serves to maintain important social and community relationships in both symbolic and concrete forms. The case of the wind farm however shows the complex nature of debates surrounding production and development and the difficulties of capturing communal benefit from such projects. Some of the networks that are involved in the wind farm debate have been described here, and they illustrate the often fragmentary qualities of rural social life. We therefore need to be wary of using simplistic notions of 'community' or 'village' and so on, because of the variety of interests and groups that are likely to be present. As described by Williams (1973) and Newby (1979) in an English context, this has long been the case, but given the propensity of development agencies to use 'the community' (in a uniform and undifferentiated sense) as a central symbol in rural development, it seems more important than ever that this is recognised.

We have also explored some of the ways in which rural areas in Europe are using marketing and branding to promote themselves and products from their area. Christopher Ray has argued that such activities can lead to the growth of a stronger identity within rural areas that itself is a process which is likely to have positive outcomes for development in terms of social capital. We have looked at this in terms of how environments are constructed through such branding and commodifications. By considering the diversity of networks that exist within and involve rural areas, we can provide a critique of these processes. In theoretical terms this would begin to establish the importance of 'place' in studies of social capital and networks, whereby the way that the specific locality is understood by residents and others will be important for the way that social capital is mobilised and development itself takes place.

Again, the case of the wind farm in Skye and Lochalsh illustrates the relationships between perception of the environment, identity and development that are played out as particular projects are put forward. There, naturalistic images of the landscape in marketing may subtly preclude alternative ways of understanding the environment – such as a worked, productive landscape. In a similar way, the marketing of Dalhalla concert venue in Dalarna in Sweden may not acknowledge the labour of local people in making the quarry, and yet the venue relies on continuing local labour through voluntary work. In this case, the venue has been a success in mobilising a localised sense of community, although the tensions between local and national may be ongoing. But a mechanism has been found to translate some of the value that the 'national' society finds in Dalhalla into a local level community benefit, through the village association system.

In summary, there are cases where networks produce certain representations of environments that are tied to particular activities. Understanding such networks is an approach that differs to standard depictions of socio-economic change in rural areas (such as between productivist and post-productivist eras), and it may more accurately convey the fragmentary and contested nature of contemporary rural social and economic change in Europe. The key issues that emerge are:

- the complexity of relationships between agriculture and rural development processes, which now more than ever involve issues of imagery, landscape and identity;
- the relationship between local and national scales of networking and development;
- the capturing of appropriate benefits from new uses of environments that may not have a direct market value.

   Underlying these factors are the existing patterns of land ownership and use that will inevitably inform responses to any new initiatives in regard to rural environments. In much of Scotland, landlords still exert significant influence over local environments, although new forms of community and not-for-profit communal ownership are beginning to occur. In the Nordic countries, the tradition of *allemansrätten* can affect how environments are perceived as a resource that can be privatised. The lesson is that these areas have continuing social structures, despite the changes to policy and economic circumstances, which need to be understood and valued appropriately. A balance needs to be struck between supporting and maintaining those elements that contribute positively to the goals for rural development and sustainable livelihoods as defined by the various local communities, and allowing for change where those areas have a chance to undo the 'peripheralising' tendencies of the national and supra-national political and economic structures of Europe, and their associated discourses of rurality.

Chapter 8
Conclusions: Comparing Rural Development

Jo Vergunst and Mark Shucksmith

This final chapter draws together the threads that have been followed in this book.[1] Although the research brief has been wide, here we summarise the principal lessons and outline the benefits of examining rural development from the perspective of social capital and networks. But rather than defining the central tenets of social capital at the outset, we have preferred to explore how different social dynamics in rural Europe can exist alongside each other. Various aspects of the concept of social capital can be used to describe them. As a result, our 'story' of social capital is one of variety, and the lesson is that social capital is not a single solution to development problems. We have found that social capital is best understood as a metaphor for the qualities of some social relationships that allow other benefits to be secured through them. An awareness of social capital is thus an awareness of the *process* of development. It is at this conceptual level that these conclusions are, for the most part, pitched. The qualitative methodologies adopted by the research teams have enabled a depth of explanation for how remote rural areas are dealing with the changing political, economic and social circumstances that they face.

Social capital and networks help analytically to characterise the social and economic relationships that people have with each other, and to understand the importance of processual aspects of development. To investigate the role of social capital, as opposed to simply documenting its presence, it has been fruitful to focus on networks. We have understood networks as articulating the flows of information and resources that produce rural development, and, indeed, society more generally. Focusing on networks therefore allows us to investigate the mechanisms by which people capture or contain benefits of development, and also illustrates the ways in which people in rural areas are becoming linked in complex ways to different scales of economy and society.

This chapter starts by reflecting on the process of maintaining an analytical unity through the variety of case studies and approaches present in this research. A consideration of rural governance structures follows, and we then draw out the theme of how development can be seen as a process of continuity and change in social capital and networks. We ask how valued aspects of local social and economic life can be maintained, while allowing for positive and appropriate

---

1 Some parts of this chapter draw on Lee [Vergunst] et al. (2005).

development. Subsequently we argue that heterogeneity and inclusiveness are key criteria for successful social capital and networks, and these are examined in the context of 'visibility' and social identity in rural Europe. We suggest that contestation over forms of identity in rural Europe is one of the key processes in how social capital and networks are operating in development. Finally, we make the case that the analysis presented here can contribute to the reformulation of models of endogenous and exogenous development, and we point to the ways in which policy can incorporate and encourage such development.

Stories of Flexible Rural Development Research

Recent research has focused on the need to identify and value new kinds of local resources in ways that can be used in rural development. As outlined in the introduction, ideas about social capital can be used to draw attention to some of the non-economic contexts for development. Bryden and Hart, within the EU project *Dynamics of Rural Areas* (DORA), describe the importance of 'non-tangible' factors in development, including local culture and the quality of associations in community, networks and institutions. They write that: 'One conclusion of DORA, therefore, is that collective cultural and social features of some local areas, often (but not always) rooted in a long-established pattern of shared history, played a leading role in explaining Differential Economic Performance (DEP) for half the regions chosen by this study' (Bryden and Hart 2001, 5). In this project we have started from the assumption that there will be aspects of associational life in all areas that affect the process of development. We have taken on the specific research topic of social capital in order to investigate its utility in both understanding the trajectory of current development and in setting out the circumstances in which development can be undertaken in the future. This is perhaps a broader remit than the explanation of differences in economic performances, and our material can be used to argue that economic performance is only one part of a positive development outcome. Social inclusion, service provision, quality of life and the environment are also criteria that are important to people living in rural areas.

However, while there has been broad agreement within the research teams on the importance of social factors, the particular characterisation of these as a 'capital' has varied between teams and the contexts that they have worked in. Some comments on our own story of getting to grips with social capital will illustrate the links between theory and data in our research and also begin the description of diversity in European rural development practices.

In the introduction, we described how different understandings of social capital are based in broader views of society and social process, and these of course have deep roots in social science. Where Durkheim viewed society with the idea of function in mind, in order to see how social groups maintain themselves as a coherent whole, Weber and Marx were concerned with the ways that individuals and segmented social groups act in relation to each other. Although Marx focused on differences in socio-economic class in terms of relationships to the means of production, distinctions are now understood to also exist in regard to gender, ethnicity, age and many other forms of identity. These variations in concepts of society also underlie differences in how social capital can be described, as discussed in the introduction. Putnam's social capital relates to norms of trust and reciprocity that can contribute to a 'functioning' society (Putnam 1993). It is understood to belong to the community as a whole. Bourdieu (1984) emphasises the relational aspect of social capital, and how it that can be articulated through differential power relations that include social class.

Our research teams have followed a similar diversity of approaches to social capital. We have refrained from imposing a single, all-encompassing definition of social capital, partly so that that the researchers in this large project could follow their own theoretical interests, but also, and perhaps more significantly, so that the differences between the research areas and themes could be explored in a more sensitive way. For example, Cecchi (Chapter 3) writes that 'social capital is the result of the use of resources – that might have been used in a different way – whose benefits can influence the performance of the community for a long period of time.' This frames his subsequent interest in the relationship between the provision of public services and local social structures. Social capital is understood as a phenomenon that can be invested in, as public services can provide a model for collective action that can be taken on by local actors. The collective aspects to social capital are emphasised by this approach. Similarly, Lehto and Oksa (Chapter 2) ground their chapter in an idea of social capital as the ability to collectively exploit a set of resources, which in themselves can include group working. In the same way that Cecchi discusses how social capital is formed through the provision of services and yet is also required in the negotiation of those services, Lehto and Oksa argue that local resources are transformed through the presence of positive social relations: social capital is both a resource and a way of using other resources.

These collectivist approaches to social capital reflect a concern with the cohesiveness and 'functioning' of society in a Durkheimian sense. Lehto and Oksa, however, draw their conclusions on social capital from a detailed discussion of networking in our case study areas. As described in the introduction to this report, the extension of the concept of social capital into studies of networks has been a significant step. Networks can be seen to coalesce through relationships to particular issues. The events described by Lehto and Oksa were often of a temporary nature, bringing together people in various formations, who often had differing interests in the event. Meistad et al. (Chapter 5), Hannon and Curtis (Chapter 6), and Vergunst et al. (Chapter 7) also use networks as their key theoretical idea, describing social networks involving the environment and various ideas of identity. These 'network' accounts often result in renditions of social capital as being highly tensioned, where local development trajectories may form and split, or diverge and converge with other local, national or international paths. The ability to turn a network to a particular point of view or goal is of interest in these chapters. In doing so, the researchers use a model of society that is clearly not a neatly bounded and homogenous entity. Instead, their focus on social relations bears more in common with studies of *agency* that come from the work of Max Weber. Bourdieu's notion of social capital as incorporating the ability to exert influence over others is relevant here: as we argue in the introduction, no accumulation of capital can be disinterested.

Stenbacka and Tillberg Mattsson (Chapter 4) maintain that qualities of relationships between groups and individuals can be important in social capital. They perhaps come closest to reconciling the two approaches to social capital outlined here, arguing that while active voluntary associations can contribute as a whole to development processes, connections between actors on an individual level need to be understood in order to ascertain specific effects on different social groups. Age, gender and place-identity (in other words, perceptions of belonging to a place) are the main categories that can be factors in determining the relationship between voluntary activities and engagement in the labour market, they find. Their typology of outcomes from this relationship at the end of their chapter shows an unwillingness to prescribe a single way of understanding social capital.

This perspective on social capital goes some way to overcoming some of the difficulties with the concept that we identified in the introduction. We suggested that the under-theorised nature of social capital means that it is hard to know exactly what is being referred to. By gathering qualitative data on, broadly speaking, associational practices within the case study areas, we have produced a set of specific accounts that are grounded within wider social science theory. That the accounts differ in their particular rendition of social capital reflects in equal measure the empirical diversity of the case study material and the theoretical interests of the researchers, both individually and as a group. The choice of themes in the project, upon which each chapter is based, came about through the relating of the empirical to the theoretical concepts – a flexible process that did not happen once and for all, but was ongoing through the life of the research. This is why we characterise our approaches to social capital as stories, to repeat from our introduction: narrative explorations of process, rather than cause-and-effect explanations of difference.

Governance and Democracy as a Basis for Social Capital

Drawing more specific conclusions is a matter of tracking back to the case studies, and identifying the most salient cross-cutting issues. Underlying much of what was going on were the structures of governance in the study areas, which, it turns out, are remarkably diverse even within the five out of six cases that are within the EU. While the EU structures are of course shared and the national governments share certain similarities, the local structures of government are generally very different to each other. To start with, the difference between areas with a relatively strong local municipality form of government, present in the Nordic areas, and the much more centralised systems of Scotland and Ireland is easy to recognise. Cecchi presents the Italian case as being in some ways intermediate, as government exists on a similarly local level to the Nordic areas but has not been able to take a lead role in development. In his discussion of social capital and public service provision, Cecchi describes the potential impact of increasing services as a catalyst for further local collective activities.

The contribution of a strong municipality to development was underlined in the Finnish case study. The Finnish case shows how the municipality of Sotkamo shifted in the 1990s from being primarily a provider of welfare services to being an important development agency in its own right. It was able to found enterprises that were perceived to be of strategic importance (such as a hotel and skiing facilities) but then sell them as private businesses once they were running. This model has proved very successful and the municipality has been widely recognised for its contributions. Strong networks between public and private sectors have been the basis for good performance.

Within this, the willingness of the municipality and other actors to take risks in development strategies can be identified as significant. Also in Finland, Lehto and Oksa describes the failure of a milk processing initiative that was intended to localise control of milk production and secure a greater economic return for the municipality. The response of the municipality was to look for investment opportunities in a different sector, and the expansion in tourism in the area has been the result. The authors identify the high degree of trust between people involved in development, which allows them to continue working together despite occasional conflicts and failures. The ability to see a 'failure' as a collective learning experience can be said to be derived from and contribute to the further creation of social capital in the area. Emphasis is on the specific processes of networking between municipal and private sectors (again based in a strong participative local democracy) and the quality of networking between specific participants that allowed for learning to take place. We might contrast this with the long and painful process of deciding whether to build a wind farm on the island of Skye in the Scottish case study, where the constant shift between 'local' (although really regional) and national government inquiries, decisions and appeals has taken many years.

At a time of reform of welfare state economies in many parts of Europe, novel roles for local government may hint at solutions. If local government can both take on new responsibilities and diversify its own activities, in the form of new networks, a more participative approach to both service provision and rural development more generally could be envisaged. Basing this in democratic, close-at-hand governance as a flexible rather than a fixed process should help rural inhabitants to maintain their stake in development.

Development from Networks: Continuities in Rural Europe

Discussing the mobilisation of networks raises issues of how both pre-existing and new networks are used in rural development. Theorisations of social capital within the development literature are often vague on the relationship between existing social structures and the ability to create new patterns of action. If we understand social capital to refer primarily to social norms, we might emphasise the current structures and behaviours in an area that are likely to lead to positive development. Alternatively, we could look more at the creation of social capital and new networks, usually as a precursor to economic development. Using the material presented in this report, the relationship between these 'roles' of social capital can probed in more detail.

The dynamic between the pre-existing and the new networks for development is clearly significant. Firstly, we should note the importance for rural development of the many continuing social and socio-economic networks that make up life in rural Europe. Secondly, networks that are formed specifically for development give a way into understanding how development is planned and occurs as a specific process, and one topic of particular relevance here is the mechanisms for communicating between various public and private spheres. The overall lesson is that aspects of continuity in networks and social capital should be stressed as much as the changes that are often central to models of development. We will look at each in turn here.

There are many types of network in our case study areas that were not planned as specific development actions but nonetheless make valuable contributions to social and economic life. An important conclusion is that many people in rural Europe have a clear idea of the activities and senses of identity that are particular to their area and give meaning to it. The attitudes that are held towards them can be far from discourses of 'development', in that people are often concerned to maintain the viability and vitality of their existing ways of life, rather than radically alter them. There is a potential and often realised conflict here between different attitudes towards the development process itself.

Some of the networks described in this book can be used to illustrate this point. Stenbacka and Tillberg Mattsson (Chapter 4) show how networks related to sport and music in Sweden are a focus for more than just the activities themselves. They provide occasions for socialising and constructing a sense of identity with a place. They also add a richness to local life that is not quantifiable in narrow economic terms and yet does affect the way that people take part in the local society and economy. As such, they appear to fit well with Putnamian characterisations of social capital, in that people are able to find benefits in membership of the groups that do not directly relate merely to sport and music. But it is apparent that the opportunities for socio-economic engagements (rather than solely social or cultural) are highly variable, as are the purposes to which such social interactions and senses of identity are put. Some of the associations have become gendered in distinctive ways, for example. In this case study, men are more likely to use time spent in music or sports activities for informal work or business purposes ('networking'), whereas women are less likely to seek such opportunities. Ice hockey is the dominant sport in the area, and is generally run by and for men. The perception of an increasing burden of voluntary work reported by some people also suggests that what might otherwise be seen as positive examples of social capital are complex interactions between people of varying status with different motives and access to resources. Supporting or promoting cultural traditions and heritage is by no means free of the difficulties often associated with other kinds of development activity.

The maintenance of social structures and activities was also a significant issue in other case study areas, although similar complexities were found. In Scotland, the concern to keep crofting agriculture going was related to a strong sense of identification with the heritage and lifestyle of crofting, and to the area itself where crofting is practiced. Vergunst et al. (Chapter 7) describe some of the networks that surround crofting as an activity, in which, again, there are opportunities for discussing and initiating many types of economic diversification, in addition to maintaining cohesion between those taking part in crofting. Crofting, and similarly historically and socially embedded activities in other areas, can be seen as providing a set of networking tools that can be put to many different uses by local actors. While the desire for continuity of what are often seen as integral aspects of life in an area may in itself be sufficient justification for support, these activities can also be vital for economic diversification. The food networks described in the Scottish case study are good examples of this, as are equivalent initiatives centred on food in the Swedish and Italian case studies.

In theoretical terms the importance of the 'embeddedness' of economic activity within social relationships is sustained here. Although this has been long recognised within some research circles (e.g. Granovetter 1985), it serves to draw attention to the interdependence of different kinds of development activity. It is also at the heart of most conceptualisations of social capital. However, the particular networks that articulate such social and economic relationships are significant, insofar as they involve unequal access to resources and benefits, for example. It is unrealistic to assume that the benefits of involvement in a particular network will be spread evenly between all members of it.

Many of the networks we examined are concerned with maintaining what can be understood as a 'productive landscape'. In Scotland, informants often emphasised how their livelihood and sense of identity were based in a productive relationship with the land. While 'traditional' meat production was an important part of this, new forms of food production (such as niche market seafood and horticulture) were also recognised as being significant. Furthermore, the desire for a productive landscape could also extend to changes in land use. The arguments that surrounded the negotiation over the wind farm on the common grazings in one crofting area showed how links were made between the agriculture of crofting and the potential for income from energy production. In both cases, there was a reluctance to have the land simply as a kind of visual amenity for its own sake, or just to attract tourists. There was a sense that control over production was important for the crofters, which is very different to providing a set of services for tourist consumption. However, the debates over the wind farm showed how such issues cannot be easily resolved, involving as they do many different interests.

The various development initiatives centred on food in other case study areas (examples were found in Sweden, Italy and Norway) can be understood in similar terms, as a link between the history of agricultural production and the new contexts of development in rural Europe. In Sweden the grazing associations formed to prevent the growth of bush that would block pleasing views also shows how landscape as a visual amenity and a productive entity do not need to be in opposition. Again, an activity that might not have an obvious 'development' return can affect development in a broader sense, improving the quality of life in an area. Maintaining house price values was understood to be a consequence in this case.

While the main purpose of this research has been to analyse the relevance of social capital and networks, we can consider in parallel how the continuity of attitudes and practices illustrated here may affect the notion of a post-productivist transition. Ilbery (1998) has contended that there is a shift towards practices of consumption in rural areas, whereby environmental goods, tourist services and specialist types of agriculture (such as organic) are forming a 'post-productivist' norm. Although we recognise the importance of such activities in the study areas, many of the development networks that we studied are geared towards similar productive ideals as have existed in rural Europe over a longer time span. We thus concur with Wilson (2001), who argues that despite the lessening role of agriculture itself in rural economies, productivist ways of thinking continue to be central in how many farmers construct their identities.

In addition, however, productivist ideals can contribute to more than just agricultural forms of development. Many rural communities have been successful in diversifying their economies in such a way that does not cut them off from agricultural practices. Food networks can be one strategy amongst a range of economic activities (such as tourism and other service provision) that are now more likely to underpin successful rural areas. This is in keeping with Marsden's description of a 'rural development' sustainability paradigm that may be emerging in Europe (Marsden 2003). He argues that rural development goals should include local initiative and control over development strategies, in distinction to the urban-centred 'amenity' demands of post-productivism. The concern we have found in the case study areas with building on existing ways of life very much backs up this perspective.

Networks for Development: Planned Social Change

In this section we discuss some of the networks in the case study areas that have been founded in recent years specifically as development efforts. We cross from a consideration of social capital as set of pre-existing or traditional social relationships into how social capital can be created in new ways of working. Our focus on the networks that people engage in begins to illuminate what happens during such initiatives. The comparative perspective adopted in the research also draws attention to both the diversity and commonalities that are present in rural Europe.

The initiatives in the case study areas of Norway and Ireland provide good examples of attempts to use new kinds of social relationships in rural development (the Mountain Region Council (MRC) and Lake District Enterprise (LDE), respectively). As described by Meistad et al. (Chapter 5), these regional partnerships have been set up in an attempt to conduct development activities on what are seen to be more appropriate scales. The institutional context of the organisations is, however, very different. The Norwegian partnership is an agglomeration of local municipalities while the Irish partnership tries to engender local control over development in what is a much more centralised national political structure. Although they both have aspirations of influencing local social and economic trends, the Norwegian partnership is a formal political structure, made up of local politicians, whereas the Irish case is a looser affiliation of politicians, business people and residents. It could be argued that the former is a top-down development initiative, concerned with more efficient service provision, while the latter is bottom-up, attempting to localise economic and social development and involve the whole community. Both share the goal of promoting regional cohesion, but from very different starting points.

One lesson to emerge here is that the qualities of a particular new network still very much depend on the pre-existing context. The 'regional' scale of networking constructed in both areas is based on very different premises, one moving from a smaller to a larger scale of local development planning, and the other *vice versa*. Through not being attached to a political structure, the Irish LDE has had to network between different sectors of the population, and in practice this has been difficult to achieve. Funding has largely come from local memberships, which have dropped in latter years. The Norwegian MRC, with a more secure structural base, has been able to carve out an appropriate role over a longer timescale. Stable funding is likely to contribute to the success of an initiative for many reasons, because in addition to the obvious practical benefits the perception that a project is only temporary in itself may downgrade the importance that people attach to it. This was also borne out in the other case study areas, in particular Scotland.

Furthermore, there is no need to presume that public-funded projects are in fact 'top-down', or that they necessarily preclude 'local' or grassroots involvement. Rather, such involvement is more likely to be related to the strength of the existing local democracy. A strong and democratic local political structure may be a more effective basis for development networks which can then involve business and other residential interests. The alternative, a *laissez-faire* approach where businesses and the civic sector is expected to take the lead in development initiatives, is hard to sustain in the medium and long term.

There are however still some notable similarities between the Irish and Norwegian cases. The Mountain Regional Council has been successful in some practical measures in industrial development planning and the provision of decentralised university courses on the level of the region. The Irish Lake District Enterprise, meanwhile, has also been able to provide business support and other forms of training for local people. That these efforts towards training and skill provision are common to both areas provides further backing for a conclusion reached by the DORA research project, where the local availability of education and training was emphasised as a key part of successful economic performance in rural areas (Bryden and Hart 2001). Another similarity is the extent to which they try to cultivate a regional identity that was not present prior to each initiative. We return to some conclusions on the role and effect of identity in development in the following section.

We can suggest that networks planned specifically *for* development, as opposed to those that were based in older social structures and activities, have been highly variable in their effectiveness. Some demonstrate how links can fruitfully be made between business, civic and public sectors, while others show the difficulties of such partnerships and the problems of sustaining initial impetus. On the other hand, inevitably only some pre-existing social networks are likely to be able to contribute towards economic development, although those that can be in a favourable position because of their embeddedness in older social relations. But strong social networks can be part of a broader social context from which planned 'development' can emerge, even where there is no direct link with economic viability in a strict sense.

'Visibility', Rural Development and Identities

In the introduction, we described some of the links between rural development and concepts of rurality in Europe. There is a significant difference between representations of rurality as, on the one hand, a lack of the material benefits of urban living, and, on the other, as a repository of tradition, essences of culture and nature, and the nation. While the former emphasises a need for change, the latter holds out for conservation or continuity. Although they are not mutually exclusive, the difference illustrates the varying attitudes that can be brought to discussions of rurality. People involved in rural development will inevitably bring certain ways of thinking about rurality to development debates. Rural development, like rurality itself, cannot be understood as a technocratic process of linear economic and social improvement. Rather, it is always being challenged, negotiated and re-worked between people operating in networks, and who also hold different ideas about rurality (Milbourne 1997).

One of the particular processes that illustrates the importance of these representations of rurality is the marketing or branding of rural areas. An initial point to note here is that most of the case study areas we have worked in have adopted a 'logo' for use in marketing their area. Some of them – those in our Scottish, Irish and Finnish cases – bear an uncanny resemblance to each other, with an almost identical curve representing a hilly or mountainous landscape. This hints at the difficulties in creating a 'unique' territorial identity, and indeed there will inevitably be tensions when a single image is taken to stand for a diverse place.

Empirically, many of the networks for development in the case study areas were based around such marketing strategies, or 'visibility' in the more complex theorisation of Lehto and Oksa (Chapter 2). These authors argue that it is possible to organise development around the principle of visibility that involves promoting the resources and products of an area in novel ways. Some of the study areas, including Norway and Ireland, were working on becoming more visible as territorial entities. Others have already achieved a degree of visibility and face the task of turning that into successful rural development outcomes. Lehto and Oksa emphasise the importance of learning and decision-making mechanisms here. In the Finnish case, the ability to take decisions and make strategies independently of regional trends, tailored specifically to the municipality, was important in the development of recreational tourism.

The similarity of the Finnish case to the other study areas is of course partial. While all the areas have been attempting some degree of diversification of their economies away from an agricultural or forestry base, it is Sotkamo in Finland that appears to have been most successful in this regard. On the other hand, the concentration on sports and tourism in that area runs the danger of maintaining too great a reliance on particular economic sectors. Most of the efforts towards branding or marketing in the other case study areas are similarly focused on tourism. Before now, there has been relatively little critical evaluation of the effects of such branding or marketing on existing personal and social identities, or the ways that the content of branding may implicitly favour some development strategies over others. Many of the chapters in this book remedy this by engaging with the issue of identity in rural development.

The work of Christopher Ray in this regard is discussed in the Chapters 6 and 7. Ray (1998) argues that by encouraging a single and unified sense of identity in a rural area, the territory can not only be marketed to others (as a tourist destination and for the consumption of niche products), but can be marketed 'to itself' as a way of creating social capital. A further conclusion from our fieldwork is that this is a highly tensioned process that cannot be simply controlled by key development actors in a straightforward manner. People in rural Europe have much more complex senses of identity that enable membership of many different social networks. These may be based in family, occupational or territorial allegiances, for example.

The authors of Chapter 5 argue that newly-cultivated regional identities in the case study areas of Norway and Ireland are 'clothes' that can be worn for outsiders, and do not necessarily affect the way of thinking amongst inhabitants within the regions. This is a more sceptical take on the significance of such development initiatives. Nonetheless, the potential for tension between a newly constructed marketing device or territorial brand and the existing senses of identity in the case study areas has been apparent through the Norwegian, Irish and other case studies. On one level this relates to the disjuncture between territorial identities that people may hold, in that they may be unwilling to subjugate what are often more local place-identities to a region. However, it can also relate to the content of such branding strategies. As described in the Scottish case study in Chapter 7, the desire to find a neutral way of marketing an area can lead to an emphasis on the 'natural' aspects of a territory, which seems to foreclose the representation or discussion of a cultural heritage. The tendency towards such tourist development may stress the environment as visual amenity over the social and productive aspects of a place.

An exception is the Italian case study area, where local speciality products are closely tied to the history of the area. But Cecchi in Chapter 3 argues that there has been relatively little mobilisation for rural development in Maremma. The provision of services in the area is potentially a mechanism for the diversification of the economy and the development of collective activities. In practice, a low awareness of the possibilities of rural development combined with an emphasis on traditional agricultural production and subsidies has meant that these possibilities have not been realised. Although this case study illustrates the difficulties of circularity in discussions of social capital, whereby because social capital is 'missing' from an area, it cannot be 'created' for further benefits, it also describes how the provision of services in an area can encourage development of the kind of community networks that were found in other case studies, as well as bringing improvements to infrastructure.

Perhaps the most successful areas are those where networking efforts are able to overcome models of endogenous as opposed to exogenous development, or infrastructure as opposed to community development. The descriptions of social capital in this book are most relevant in describing how networks can link different sectors of development and different spatial scales. Social capital cannot be thought of as a property of closed and bounded rural communities, which merely perpetuates the myths of rurality as a preserve of old traditions. And yet, on the other hand it is very much linked to ideas of place and identity. Where social capital brings positive benefits, it is likely to be associated with a plurality of cultural identities, a mixing and interweaving of spatial scales (through, for example, diverse marketing strategies), and strong links to the multiple historical themes that characterise European rural areas.

Implications for Policy

The most fundamental conclusion emerging from this study is that social processes, through networks, are fundamental to rural development. In this sense, social capital has a vital role in rural development, along with appropriate structures of governance. The role of public policy and development agencies is to trust, foster and enable local action. This has a number of implications for policy:

'Soft rural development.' Public policy should support the social processes which are as essential to rural development as 'hard' economic intervention (in the same sense that software is as necessary as hardware to computing). In practice this means supporting rural community development – understood as an approach to working with and to building the capacity of individuals and groups within their communities. This approach seeks to strengthen communities through enhancing people's confidence, knowledge and skills, and their ability to work together. In the EU, this type of approach has been piloted successfully under the community initiative, LEADER, and it is important that this approach be maintained by DG Agriculture under the single rural development fund.

Supporting the development of vertical and horizontal networks in community action can transcend the dichotomy of endogenous/exogenous development ('bottom-up/top-down'). Issues will arise of where power and control lie in these networks, and of whose problems they are addressing and who benefits, and public bodies and development agencies should be alert to these aspects when offering support and when working with voluntary and community bodies. Training of local and regional officials in the social processes surrounding local development is crucial.

In offering grants and other support, development agencies should prioritise collective action which is both inclusive and reflexive, and should support new arenas for interaction. Good networks are inclusive, facilitating collective learning, allowing sharing of success and generating wider social acceptance. In this context, it is notable that most expenditure under the EU Rural Development Regulation is targeted at individuals rather than collective activities. There is scope for the RDR to be more effective through promoting collective action.

Maintain and build on the social structures that already exist, rather than offering support only for new structures and projects. New projects may compete with existing structures and social networks, causing duplication, displacement and conflict, and then face continuing crises of funding and of legitimacy. Where new structures are created it is important that these issues are anticipated. Local practices should be the starting points for development, forming continuities with the past, defining strengths and weaknesses, and forming the basis for action.

Value alternative development discourses and promote constructive dialogue in the criteria and conditions for public funding. It would be very surprising if a 'community' had only one uniform view, and indeed our studies found evidence of multiple voices in each area, often each laying claim to being the authentic view of the whole community. Funders should require evidence that applicants have sought to include alternative voices in their proposals.

This comment also applies to the branding and marketing of rural places. It is important to base branding on a plurality of cultural identities and to link this to cultures of everyday life through a broad participative process. Newly constructed regional identities will only succeed in mobilising common efforts towards shared objectives where these supplement and build on multiple local identities.

Appropriate structures of governance are also essential to facilitate local leadership and innovation, as noted in several previous studies. Rural areas and people require strong support from national government and the EU, as well as from regional agencies and the private sector, and it is essential that these set a coherent framework within which participative local development initiatives can flourish. Within such a framework, rural development can be pursued which is locally embedded, socially inclusive and, often, producing or encompassing networks which link social scales. Successful development of this type frees rural areas from stereotypes of backwardness, remoteness and parochialism, and yet allows them to retain control of distinctive and valued cultural and environmental features, with long-term beneficial results.

Finally, talk of a post-productive countryside is (still) premature, and sometimes inappropriate. For many people, production – of food or other commodities, but also of particular forms of social relationships, landscapes, land use and even kinds of people – is still important. Moreover, the environment in our rural communities is understood as not just as the plants, soil and animals, but also as relating to work, heritage and land tenure, and other social and cultural forms. The idea of multifunctionality in agriculture needs to be broad enough to include such themes.

References

Abram, S. and Waldren, J. (1998), *Anthropological Perspectives on Local Development: Knowledge and Sentiments in Conflict* (London Routledge).

Almås, R. (2003), 'Two roads to the global village: a comparison of how a coastal and a mountain region of Norway have found strategies to cope with globalization', in R. Almås and G. Lawrence (eds) *Globalisation, Localization and Sustainable Livelihoods* (Aldershot: Ashgate).

Almås, R. (1985), *Teoretiske Perspektiver Omkring Temaet 'Lokalsamfunnsretta Tiltaksarbeid'* (Trondheim: IFIM, Sintef/Bygdeforskning, Universitetet i Trondheim).

Árnason, A., Lee, J. and Nightingale, A. (2003), *Crofting Diversification: Networks and Rural Development in Skye and Lochalsh, Scotland*, Scottish National Report. Restructuring in Marginal Rural Areas (RESTRIM) project (Arkleton Centre for Rural Development Research, University of Aberdeen).

Arrow, K. (2000), 'Observations on social capital', in P. Dasgupta and I. Serageldin (eds) *Social Capital: A Multifaceted Perspective* (Washington, DC: World Bank).

Barker, D.K. (2000), 'Dualisms, discourse, and development', in U. Narayan and S. Harding (eds) *Decentering the Center: Philosophy for a Multicultural, Postcolonial, and Feminist World* (Indianapolis: Indiana University Press).

Baron, S., Field, J. and Schuller, T. (2000), *Social Capital: Critical Perspectives* (Oxford: Oxford University Press).

Basile, E. and Cecchi, C. (2001), *La Trasformazione Post-Industriale della Campagna. Dall'agricoltura ai Sistemi Locali Rurali* (Torino: Rosenberg and Sellier).

Bassand, M. (1986), 'The socio cultural dimension of self-reliant development', in M. Bassand, M. Brugger, J. Bryden, J. Friedmann, and B. Stuckey (eds) *Self-reliant Development in Europe* (Aldershot: Gower).

Becattini, G. (1999), 'Introduzione', in G. Becattini (ed.) *Modelli Locali di Sviluppo* (Bologna: Il Mulino).

Becattini, G. (ed.) (1987), *Mercato e Forze Locali* (Bologna: Il Mulino).

Berglund, A. (1998), *Lokala Utvecklingsgrupper på Landsbygden. Analys av Några Lokala Utvecklingsgrupper i Termer av Platsrelaterad Gemenskap, Platsrelaterad Social Rörelse och Systemintegrerad Lokal Organisation*, Akademisk avhandling. Geografiska regionstudier nr. 38 (Kulturgeografiska Institutionen, Uppsala Universitet).

Bianchi, P. (1998), *Industrial Policies and Economic Integration. Learning from European Experiences* (Routledge: London).

Bourdieu, P. (1991), *Language and Symbolic Power* (Cambridge, MA: Harvard University Press).

Bourdieu, P. (1984 [1979]), *Distinction* (London: Routledge and Kegan Paul).

Bourdieu, P. (1980), 'Le capital social: notes provisoires'. *Actes Recherches: Science Social* 31:2, 2–3.

Bourdieu, P. and Wacquant, L. (1992), *An Invitation to Reflexive Sociology* (Cambridge: Polity Press).

Braun, B. and Castree, N. (eds) (1998), *Remaking Reality. Nature at the Millennium* (London: Routledge).

Bruckmeier, K. (ed.) (2001), *Innovating Rural Evaluation. Social Science in the Interdisciplinary Evaluation of Rural Development* (Berlin/Freiburg: INFIS).

Bryden, J. and Hart, K. (2004), *A New Approach to Rural Development in Europe – Germany, Greece, Scotland and Sweden* (Ceredigion: Edward Mellen).

Callendar, R. (1998), *How Scotland is Owned* (Edinburgh: Canongate).

Callon, M. (1986), 'Some elements of a sociology of translation', in J. Law (ed.) *Power, Action, Belief: A New Sociology of Knowledge* (London: Routledge and Kegan Paul).

Canavan, T. (1993), 'Locality and divided histories', in P. O' Drisceoil (ed.) *Regions: Identity and Power* (Belfast: Institute of Irish Studies).

Castree, N. and Braun, B. (eds) (2001), *Social Nature. Theory, Practice and Politics* (Oxford: Blackwell).

Caxton, W. (1988 [1840]), *The Description of Britain*, ed. M. Collins (New York: Weidenfeld and Nicolson).

Cecchi, C. (2002), 'Sistemi locali rurali, e aree di specializzazione agricola', in E. Basile and D. Romano (eds) *Sviluppo Rurale: Società, Territorio, Impresa* (Milano: Franco Angeli).

Cecchi, C. (2001), *Rural Development and Local Systems. The Case of the 'Maremma Rural District* (Department of City and Regional Planning, University of Wales, College of Cardiff).

Cecchi, C. and Micocci, A. (2003), *Public and Private Services: Building up Social Capital in Rural Marginal Areas. The Remoteness of Maremma, a Rural European District*, National Report – Italy. Restructuring in Marginal Rural Areas (RESTRIM) project (Dipartimento di Economia Pubblica, Università degli Studi di Roma 'La Sapienza', Roma, Italy).

Cecchi, C. and Micocci, A. (2002), *Hot Springs, the Sileno and Malaria. Rurality as a Choice*, Italian context report, Restructuring in Marginal Rural Areas (RESTRIM). (Dipartimento di Economia Pubblica, Università degli Studi di Roma 'La Sapienza', Roma, Italy).

Clark, H. (1994), 'Sites of resistance: place, "race" and gender as sources of empowerment', in P. Jackson and J. Penrose (eds) *Constructions of Race, Place and Nation* (Minneapolis: University of Minnesota Press).

Clark, H. (1985), *The Symbolic Construction of Community* (London: Tavistock).

Cohen, A. (1987), *Whalsay. Symbol, Segment and Boundary in a Shetland Island Community* (Manchester: Manchester University Press).

Colby, K. (1988), 'Public access to private land – *allemansrätt* in Sweden', *Landscape and Urban Planning* 15, 253–64.

Coleman, J. (1990), *Foundations of Social Theory* (Cambridge and London: Harvard University Press).

Crocker, J., Potapchuk, W., and Schechte, W. (1998), *Systems Reform and Local Government: Improving Outcomes for Children, Families and Neighbourhoods* (Washington, DC: Programme for Community Problem Solving).

Descola, P. and Pálsson, G. (eds) (1996), *Nature and Society. Anthropological Perspectives* (London: Routledge).

Eriksen, T. (1997), 'The nation as human being – a metaphor in a mid-life crisis? Notes on the imminent collapse of Norwegian national identity', in K.F. Olwig and K. Hastrup (eds) *Siting Culture: The Shifting Anthropological Object* (London: Routledge).

Escobar, A. (1995), *Encountering Development: The Making and Unmaking of the Third World* (Princeton, NJ: Princeton University Press.).

Euresco (2001), *Conference on Social Capital: Interdisciplinary Perspectives*, Exeter.

European Commission (1988), *The Future of Rural Society. COM (88) 501* (Brussels: EC).

Eurostat (2003), *50 Years of Figures on Europe. Data 1952–2001* (Office for Official Publications of the European Communities: Luxembourg).

Falk, I. and Kilpatrick, S. (2000), 'What *is* social capital? A study of interaction in a rural community', *Sociologia Ruralis* 40:1, 87–110.

Farner, A. (1988), *Usynlig Arbeid og Skjulte Ressurser – Ressurser for og i Lokalt Arbeid*, Report no. 15 (Norwegian Institute for Urban and Regional Research, Oslo).

Ferguson, J. (1997), 'Paradoxes of sovereignty and independence: "real" and "pseudo" nation-states and the depoliticization of poverty', in K.F. Olwig and K. Hastrup (eds) *Siting Culture: the Shifting Anthropological Object* (London: Routledge).

Ferrara, M. (1998), *Targeting Welfare in a 'Soft' State. Italy's Winding Road to Selectivity*, paper prepared for the ISSA Conference, Jerusalem, 25–28 January.

Fielding, A. (1982), 'Counterurbanisation in Western Europe', *Progress in Planning* 17, 1–52.

Fine, B. (2003), 'Social capital: The World Bank's fungible friend', *Journal of Agrarian Change* 3:4, 586–603.

Fine, B. (2001), *Social Capital versus Social Theory. Political Economy and Social Science at the Turn of the Millennium* (Routledge: London).

Flora, C. and Flora, J. (1993), 'Entrepreneurial social infrastructure: A necessary ingredient', *The Annals of the Academy of Social and Political Sciences* 529, 48–58.

Flora, J. (1998), 'Social capital and communities of place,' *Rural Sociology* 63:4, 481–506.

Forberg, S. (2002), 'Finst det ein "fjellbygdidentitet"? Innleiing om kulturell identitet og utmarkforvalting på konferanse i Lom 18. januar 2002', *Utmark*, 2.

Fukuyama, F. (ed.) (1995), *Trust: The Social Virtues and the Creation of Prosperity* (New York: The Free Press).

Galjart, B. (1980), 'Counterdevelopment. A position paper', *Community Development Journal* 16, 2.

Gardner, K. and Lewis, D. (1996), *Anthropology, Development and the Post-Modern Challenge* (London: Pluto Press).

Giddens, A. (1972), 'Introduction: Durkheim's writings in sociology and social philosophy', in A. Giddens (ed.) *Emile Durkheim: Selected Writings* (Cambridge: Cambridge University Press).

Gilbert, A. (1988), 'The new regional geography in English and French-speaking countries', *Progress in Human Geography* 12, 208–28.

Gilbert, M. (1997), 'Identity, space, and politics: a critique of the poverty debates', in J. Jones, H. Nast and S. Roberts (eds) *Thresholds in Feminist Geography: Difference, Methodology, Representation* (Lanham, MD: Rowman and Littlefield).

Glaeser, E. (2001), 'The formation of social capital', *Canadian Journal of Public Policy* 2, 34–40.

Granovetter, M. (1985), 'Economic action and social structure: The problem of embeddedness', *American Journal of Sociology*, 91, 481–510.

Grootaert, C. and van Bastelaer, T. (2002), 'Social capital: from definition to measurement', in C. Grootaert and T. van Bastelaer (eds) *Understanding and Measuring Social Capital. A Multidisciplinary Tool for Practitioners* (Washington DC: The World Bank).

Hall, C. and Page, S. (1999), *The Geography of Tourism and Recreation: Environment, Place and Space* (London and New York: Routledge).

Hall, S. (1990), 'Cultural identity and diaspora', in J. Rutherford (ed.) *Identity* (London: Lawrence and Wishart).

Hannon, F., Kinlen, L. and Curtin, C. (2003), *Creating an Irish Lake District to Accelerate Rural Development*, National Report (The Lake District, County Mayo, Ireland). Restructuring in Marginal Rural Areas (RESTRIM) project. (Department of Political Science and Sociology, NUI Galway).

Hansen, K. (1999), 'Emerging ethnification in marginal areas of Sweden', *Sociologia Ruralis* 39:3, 294–310.

Haraway, D. (1997), '"Gender" for a Marxist dictionary: the sexual politics of a word', in L. McDowell and J. Sharp (eds) *Space, Gender, Knowledge. Feminist Readings* (London: Arnold).

Hatch, E. (1973), *Theories of Man and Culture* (New York: Columbia University Press).

Herlitz, U. (1998), *Bygderörelsen i Sverige* (Östersund: Swedish Institute for Regional Research).

Hillmert S. (2003), *Welfare State Regimes and Life-Course Patterns: An Introduction*, http://www.mpib-berlin.mpg.de/en/institut/dok/full/e2001.0232/frames/paper.htm.

Hirsch, E. and O'Hanlon, M. (eds) (1995), *The Anthropology of Landscape* (Oxford: Clarendon Press).

Hoggart, K., Buller, H. and Black, R. (1995), *Rural Europe. Identity and Change* (London: Arnold).

Hoggart, K. and Paniagua, A. (2001), 'What rural restructuring?', *Journal of Rural Studies* 17, 41–62.

Honderich, T. (ed.) (1995), *The Oxford Companion to Philosophy* (Oxford: Oxford University Press).

Hunter, J. (1976), *The Making of the Crofting Community* (Edinburgh: Donald).

Ilbery, B. (ed.) (1998), *The Geography of Rural Social Change* (Harlow: Longman).

Jackson, P. and Penrose, J. (1994), 'Introduction', in P. Jackson and J. Penrose (eds) *Constructions of Race, Place and Nation* (Minneapolis: University of Minnesota Press).

Jeppson Grassman, E. (1997), *För Andra och för Mig: Det Frivilliga Arbetets Innebörder* (Sköndalsinstitutets skriftserie 8. Sköndalsinstitutet).

Jeppson Grassman, E. and Svedberg, L. (1999), 'Medborgarskapets gestaltningar: insatser i och utanför föreningslivet', *Statens Offentliga Utredningar 1999: 84* (Civilsamhället. Demokratiutredningens forskarvolym).

Jeppson Grassman, E. and Svedberg, L. (1996), 'Voluntary action in a Scandinavian welfare context: the case of Sweden', *Nonprofit and Voluntary Sector Quarterly* 25:3, 415–27.

Johannisson, B. (2002), *The Sparse Nordic Settlements in the Global Economy: Challenges for a New Generation of Entrepreneurs*, paper presented at the conference 'Sparsely Populated Regions in the Global Economy', Stockholm, Sweden, 14–15 October.

Jonsson I., Rydén G., Tillberg K. (2002), *Innovativeness and Traditionalism: Rural Development in Leksand and Rättvik, Sweden*, Swedish context report, Restructuring in Marginal Rural Areas (RESTRIM) (Dalarnas forskningsråd (DFR), Dalarna Research Institute, Falun, Sweden).

Jonsson, I. Rydén, G. Tillberg, K. (2001), *The Role of Social Capital in Rural Development*, Swedish Context Report for the EU research project Restructuring in Marginal Rural Areas (RESTRIM) (Dalarnas forskningsråd (DFR), Dalarna Research Institute, Falun, Sweden).

Kendal J. (2000), 'The mainstreaming of the third sector into public policy in England in the late 1990's: whys and wherefores', *Policy and Politics* 28:4, 541–62.

Kinlen L. (2002a), *The Lake District, County Mayo*, Irish Context Report, Restructuring in Marginal Rural Areas (RESTRIM) (The International Centre for Development Studies, National University of Ireland, Galway, Ireland).

Kinlen L. (2002b), *RESTRIM Project Analysis of Questionnaire*, Irish questionnaire report, Restructuring in Marginal Rural Areas (RESTRIM) (The International Centre for Development Studies, National University of Ireland, Galway, Ireland).

Kolamo, S. (1998), 'Urheilu, Maantiede ja Paikan Tunne' ['Sports, Geography and the Sense of Place'], unpublished Master's thesis, Esimerkkinä Sotkamon Jymyn pesäpallomenestys. University of Oulu, Department of Geography.

Kovacs, I. and Kucerova, E. (2006), 'The project class in central Europe', *Sociologia Ruralis* 46:1, 3–21.

Krishna, A. and Shrader, E. (1999), *Social Capital and Poverty Reduction*, Conference on Social Capital and Poverty Reduction, Washington.

Latour, B. (1986), 'The power of association', in J. Law (ed.) *Power, Action, Belief: A New Sociology of Knowledge* (London: Routledge and Kegan Paul).

Lee [Vergunst], J. (2007), 'Experiencing landscape: Orkney hill land and farming', *Journal of Rural Studies* 23:1, 88–100.

Lee [Vergunst], J. and Árnason, A. (2003), *Questionnaire Report for Skye and Lochalsh*, Scottish questionnaire report, Restructuring in Marginal Rural Areas (RESTRIM) (Arkleton Centre For Rural Development Research, University Of Aberdeen, Scotland).

Lee [Vergunst], J., Árnason, A., Nightingale, A. and Shucksmith, M. (2005), 'Networking: Social capital and identities in European rural development', *Sociologia Ruralis* 45:4, 269–83.

Lehto, E. (2002), *Coping Strategies and Regional Policies – Social Capital in the Nordic Peripheries – Finland*, NordRegio Working Paper 2002: 7, Stockholm.

Lehto, E. and Oksa, J. (2003), *The Case of Sotkamo: Social Capital and Important Events in Local Development*, National Report, Finland, Restructuring in Marginal Rural Areas (RESTRIM) (Research and Development Centre of Kajaani, University of Oulu and University of Joensuu, Finland).

Lehto, E. and Oksa, J. (2002a), *Rural Development in Kainuu Region in Finland: A Case of Sotkamo*, Finnish Context Report, Restructuring in Marginal Rural Areas (RESTRIM) (Research and Development Centre of Kajaani, University of Oulu and University of Joensuu, Finland).

Lehto, E. and Oksa, J. (2002b), *RESTRIM Questionnaire Study of Sotkamo, Finland*, Finnish Questionnaire Report, Restructuring in Marginal Rural Areas (RESTRIM) (Research and Development Centre of Kajaani, University of Oulu and University of Joensuu, Finland).

Little, J. (1997), 'Construction of rural women's voluntary work', *Gender, Place and Culture* 4:2, 197–209.

Lowe, P., Ray, C., Ward, N., Wood, D., and Woodward, R. (1998), *Participation in Rural Development: A Review of European Experience* (Centre for Rural Economy research report, Newcastle upon Tyne).

Lowndes, V. (2000), Women and social capital: a comment on Hall's 'Social Capital in Britain', *British Journal of Political Science* 30, 533–40.

Lundström, T. and Wijkström, F. (1997), *The Nonprofit Sector in Sweden* (Manchester: Manchester University Press).

Lysgård, H. (2001), *Produksjon av Rom og Identitet i Transnasjonale Regioner. Et Eeksempel fra det Politiske Samarbeidet i Nidt-Norden* (Dr. polit.-avhandling, Geografisk institutt, NTNU Trondheim).

McDonagh, J. (2001), *Renegotiating Rural Development in Ireland* (Aldershot: Ashgate).

McDowell, L. (1999), *Gender, Identity and Place: Understanding Feminist Geographies* (Minneapolis: University of Minnesota Press).

Macaulay, T. (1986 [1848–61]), *The History of England*, ed. H. Trevor-Roper (London: Penguin).

Mackenzie, C. (2004), 'Policy entrepreneurship in Australia: a conceptual review and application', *Australian Journal of Political Science* 39:2, 367–86.

Marsden, T. (2003), *The Condition of Rural Sustainability* (Assen: Royal Van Gorcum).

Marsden, T. (1998), 'Economic perspectives', in B. Ilbery (ed.) *Geography of Rural Social Change* (Harlow: Longman).

Marsden, T. and Murdoch, J. (1998), 'Editorial: The shifting nature of rural governance and community participation', *Journal of Rural Studies* 14:1, 1–4.

Massey, D. and Jess, M. (1995), *A Place in the World: Places, Culture and Globalisation* (Oxford: Oxford University Press and Open University Press).

Meistad, T. (2001), *Municipal Slogans and Emblems – Subjective Local Identities Used Strategically in Local Community Development* (Rural Transfer Network, University of Aberdeen).

Micocci A. (2002), *Questionnaire Analysis: The Case of Italy*, Italian questionnaire report, Restructuring in Marginal Rural Areas (RESTRIM) (Dipartimento di Economia Pubblica, Università degli Studi di Roma 'La Sapienza', Roma, Italy).

Milbourne, P. (ed.) (1997), *Revealing Rural 'Others': Representation, Power and Identity in the British Countryside* (London: Pinter).

Milne, D. (1997), 'Placeless power: constitutionalism confronts peripherality', *North* 8.

Morris-Suzuki, T. (1998), *Re-inventing Japan. Time, Space, Nation* (London: M.E. Sharpe).

Murdoch, J. (2000). 'Networks: A new paradigm for rural development?', *Journal of Rural Studies* 16, 407–19.

Murdoch, J. and Abram, S. (1998), 'Defining the limits of rural governance', *Journal of Rural Studies* 14:1, 41–50.

Narayan, D. (1999), *Bonds and Bridges: Social Capital and Poverty* (Washington DC: World Bank).

Nelson, N. and Wright, S. (1995), *Power and Participatory Development: Theory and Practice* (Intermediate Technology Publications).

Nightingale, A. (2002), *The Magic of Skye: Innovation, Growth and Rurality in Skye and Lochalsh*, Scottish context report, Restructuring in Marginal Rural Areas (RESTRIM) (Arkleton Centre For Rural Development Research, University Of Aberdeen, Scotland).

Office for National Statistics (2001), *Social Capital. A Review of the Literature* (Social Analysis and Reporting Division, UK Office for National Statistics).

Olwig, K. (1993), 'Sexual cosmology: Nation and landscape at the conceptual interstices of nature and culture; or, what does landscape really mean?', in B. Bender (ed.) *Landscape: Politics and Perspectives* (Oxford: Berg).

Paasi, A. (2001), 'Bounded spaces in the mobile world: deconstructing "regional identity"', *Tijdschrift voor Economische en Sociale Geografie* 93:2, 137–48.

Paasi, A. (1996), 'Regions as social and cultural constructs: reflections on recent geographical debates', in M. Idvall and A. Salomonsson (eds) *At Skapa en Region – om Identitet og Territorium* (Stockholm: nordREFO).

Paasi, A. (1986), 'The institutionalization of regions: a theoretical framework for understanding the emergence of regions and the constitution of regional identity', *Fennia* 164, 105–46.

Pagden, A. (1995), 'The effacement of difference: colonialism and the origins of nationalism in Diderot and Herder', in G. Prakask (ed.) *After Colonialism: Imperial Histories and Postcolonial Displacements* (Princeton, NJ: Princeton University Press.

Petretto, A. (1998), 'The liberalization and privatization of public utilities and the protection of user's rights: the perspective of economic theory', in M. Freedland and S. Sciarra (eds) *Public Services and Citizenship in European Law* (Oxford: Clarendon Press).

Portes, A. (1998), 'Social capital: its origins and applications in modern sociology', *Annual Review of Sociology* 24, 1–24.

Pulido, L. (1997) 'Community, place and identity', in J. Jones, H. Nast and S. Roberts (eds) *Thresholds in Feminist Geography: Difference, Methodology, Representation* (Lanham, MD: Rowman and Littlefield).

Putnam, R. (2001), 'Social capital: measurement and consequences', *Canadian Journal of Policy Research* 2:1, 41–51.

Putnam, R. (2000), *Bowling Alone. The Collapse and Revival of American Community* (Simon and Schuster: New York).

Putnam, R. (1995), 'Bowling alone: America's declining social capital', *Journal of Democracy* 6:1, 65–78.

Putnam, R. (1993), *Making Democracy Work: Civic Traditions in Modern Italy* (Princeton, NJ: Princeton University Press).

Ray, C. (2000), 'Endogenous socio-economic development in the European Union – issues of evaluation', *Journal of Rural Studies* 16, 447–58.

Ray, C. (1999), 'Endogenous development in the era of reflexive modernity', *Journal of Rural Studies* 15:3, 257–67.

Ray, C. (1998), 'Culture, intellectual property and territorial rural development' *Sociologia Ruralis* 38, 3–19.

Røiseland, A. and Aarsæther, N. (1999), 'Community and democracy: theoretical and methodological aspects of the concept of "social capital"', *Norsk Statsvitenskapelig Tidsskrift* 15, 184–201.

Rothstein, B. (2001), 'Social capital in the social democratic state. The Swedish model and civil society', *Politics and Society* 29, 289–333.

– (2000), 'Socialt kapital i den socialdemokratiska staten. Den svenska modellen och det civila samhället', *Arkiv för Studier i Arbetarrörelsens Historia* 79, 1–55.

Rowlands, J. (1995), 'Empowerment examined', *Development Practice* 5:2, 101–7.

Rye, J.F. Meistad, T and Frisvoll, S. (2003), *Regional Identity and Rural Development: The Mountain Region, Norway*, Norwegian National Report Restructuring in Marginal Rural Areas (RESTRIM) (Centre for Rural Research, Norwegian University of Science and Technology, Trondheim, Norway).

Rye, J.F. and Winge, A. (2002a), *The Mountain Region*, Norwegian context report, Restructuring in Marginal Rural Areas (RESTRIM) (Centre for Rural Research, Norwegian University of Science and Technology, Trondheim, Norway).

Rye, J.F. and Winge, A. (2002b), *A Survey of Development Priorities, Projects and Networks in the Mountain Region*, Norwegian questionnaire report, Restructuring in Marginal Rural Areas (RESTRIM) (Centre for Rural Research, Norwegian University of Science and Technology, Trondheim, Norway).

Rye, J.F. and Winge, A. (2002c), *The Role of Networks in Promoting New Usage of Farmland in the Mountain Region, Norway*, Norwegian report, Restructuring in Marginal Rural Areas (RESTRIM) (Centre for Rural Research, Norwegian University of Science and Technology, Trondheim, Norway).

Sabatini, F. (2003), *Capitale Sociale e Sviluppo Economico*, Serie Working Papers dei dottorandi (Dipartimento di Economia pubblica, Università degli studi di Roma 'La Sapienza').

Salomonsson, A. (1996), 'Regionalitet som problem', in M. Idvall and A. Salomonsson (eds) *At Skapa en Region – om Identitet og Territorium* (Stockholm: nordREFO).

Schama, S. (1996), *Landscape and Memory* (New York: Alfred A. Knopf).

Schuller, T. (2001), 'The complementary roles of human and social capital', *Canadian Journal of Policy Research*.

Schuller, T. and Baron, S. (2000), 'Social capital: a review and critique', in S. Baron, J. Field and T. Schuller (eds) *Social Capital: Critical Perspectives* (Oxford: Oxford University Press).

Scottish Executive (2003), *Skye and Lochalsh Food Van Link Group*, http:// www.scotland.gov.uk/about/ERADEN/SCU/00017108/safgrant200209. aspx (1 October 2003).

Sellar, W. (2006), 'The great land debate and the Land Reform (Scotland) Act 2003', *Norsk Geografisk Tidsskrift – Norwegian Journal of Geography* 60: 100–109.

Shortall, S. and Shucksmith, M. (1998), 'Integrated rural development: issues arising from the Scottish experience', *European Planning Studies* 6:1, 73–88.

Shucksmith, M. (2000), 'Endogenous development, social capital and social inclusion: Perspectives from LEADER in the UK', *Sociologia Ruralis* 40, 208–18.

Shucksmith, M., Thomson, K. and Roberts, D. (2005), *The CAP and the Regions: The Territorial Impact of the Common Agricultural Policy* (CABI Publishing, Wallingford).

Stark, A. and Hamrén, R. (2000), *Frivilligarbetets Kön. En Översikt* (Svenska kommunförbundet, Stockholm).

Statistics Sweden (1993), *Fritid 1976–1991. Levnadsförhållanden*, Rapport nr 85 (Stockholm: Statistics Sweden).

Stone, W. (2001), *Measuring Social Capital: Towards a Theoretically Informed Measurement Framework for Researching Social Capital in Family and Community Life* (Melbourne: Australian Institute of Family Studies).

Svendsen, G. and Svendsen, G. (2000), 'Measuring social capital: the Danish Co-operative Diary Movement', *Sociologica Ruralis*, 40:1.

Taylor, C. (1964), *The Explanation of Behaviour* (London: Routledge and Kegan Paul).

Tillberg Mattsson, K. and Stenbacka, S. (2003), *The Role of Social Capital in Local Development: The Case of Leksand and Rättvik*, National Report – Sweden, Restructuring in Marginal Rural Areas (RESTRIM) (Dalarnas forskningsråd (DFR), Dalarna Research Institute, Falun, Sweden).

Tregear, A. (2003), 'From Stilton to Vimto: Using food history to re-think typical products in rural development', *Sociologia Ruralis*, 43, 91–107.

Uslaner, E. (1999), 'Democracy and social capital', in M. Warren (ed.) *Democracy and Trust* (Cambridge: Cambridge University Press).

Väisänen, U. (1997), 'Pesäpalloilu osana elämääni' ['Finnish Baseball as part of my life]', in T. Vinha-Mustonen (ed.) *Minun Sotkamoni*, Sotkamo.

van der Ploeg, J. and Long, A. (1994), *Born From Within. Practice and Perspectives of Endogenous Development* (Assen: van Gorcum).

van 't Klooster, S., Van Asselt, M. and Koenis, S. (2002), 'Beyond the essential contestation: construction and deconstruction of regional identity', *Ethics, Place and Environment* 5:2, 109–21.

van Koppen, C. (2000), 'Resource, arcadia, lifeworld. Nature concepts in Environmental Sociology', *Sociologia Ruralis* 40, 300–18.

van Staveren, I. (2000), *A Conceptualisation of Social Capital in Economics: Commitment and Spill-Over Effects* (The Hague: Institute of Social Studies).

Veggeland, N. (1993), 'Perspektivfattig norsk regionalmelding', *NordRevy* 2.

Verdery, K. (1999), *The Political Lives of Dead Bodies. Reburial and Postsocialist Change* (New York: Columbia University Press).

Ward, N., Jackson, P. Russell, P. and Wilkinson, K. (2008), 'Productivism, post-productivism and European agricultural reform: The case of sugar', *Sociologia Ruralis* 48:2, 118–32.

Warner, M. (2001), 'Building social capital: the role of local government', *Journal of Socio-Economics* 30:2, 187–92.

Warren, M. (1999), 'Democratic theory and trust', in M. Warren (ed.) *Democracy and Trust* (Cambridge: Cambridge University Press).

Weber, M. (1978), *Economy and Society: An Outline of Interpretative Sociology*, ed. G. Roth and C. Wittich (Berkeley: University of California Press).

Whelan, K. (1996), 'The region and the intellectuals', in L. O'Dowd (ed.) *On Intellectuals and Intellectual Life in Ireland* (Belfast: Institute of Irish Studies, Queens University Belfast and the Royal Irish Academy).

Whelan, K. (1993), 'The bases of regionalism', in P. O' Drisceoil (ed.) *Regions: Identity and Power* (Belfast: The Institute of Irish Studies, Queens University Belfast and the Royal Irish Academy).

Wightman, A. (1997), *Who Owns Scotland?* (Edinburgh: Canongate).

Wijkström, Filip (1998a), *Different Faces of Civil Society* (Stockholm: Stockholm School of Economics, The Economic Research Institute).

Wijkström, Filip (1998b), 'Social ekonomi. Om mening och identitet bortom lönearbetet', in F. Wijkström (ed.) *Social ekonomi. Om kraften hos alla människor* (Utbildningsförlaget Brevskolan).

Wiklund, T. (1995), *Det Tillgjorda Landskapet* (Göteborg: Bokförlaget Korpen).

Williams, R. (1973), *The Country and the City* (London: Chatto and Windus).

Wilson, G. (2001), 'From productivism to post-productivism ... and back again? Exploring the (un)changed natural and mental landscapes of European agriculture', *Transactions of the Institute of British Geographers* N.S. 26, 77–102.

Witoszek, N. (1998), *Norske Naturmytologier. Fra Edda till Oekofilosofi* (Oslo: Pax Förlag A/S).

Wollan, G. (1994), 'Hvordan oppfattes spredt bosetting, i regionen og utenfor?', in T. Meistad (ed.) *Fire Perspektiver på Spredt Bosetting. Om den Strategiske Betydningen av Spredtbygdhet for Nord-Norge i EUs Målområde 6* (Nord-Trøndelagsforskning, rapport 17, Steinkjer).

Woods, M. (1997), 'Researching rural conflicts: hunting, local politics and actor-networks', *Journal of Rural Studies* 14, 321–40.

Woolcock, M. (1998), 'Social capital and economic development: towards a theoretical synthesis and policy framework', *Theory and Society*, 27, 151–208.

Woolcock, M. (2001), 'The place of social capital in understanding social and economic outcomes', *Canadian Journal of Policy Research* 2:1, 11°17.

World Bank. 'Social Capital. What is Social Capital?', http://web.worldbank.org/ WBSITE/EXTERNAL/TOPICS/EXTSOCIALDEVELOPMENT/EXTT SOCIALCAPITAL/0,,contentMDK:20185164~menuPK:418217~pagePK: 148956~piPK:216618~theSitePK:401015,00.html, accessed 1 April 2008.

Yin, R. (1994), *Case Study Research. Design and Methods* (London: Sage).

Index